BUILD YOUR OWN WORKING ROBOT

THE SECOND GENERATION

BUILD YOUR OWN WORKING ROBOT

THE SECOND GENERATION

DAVID L. HEISERMAN

FIRST EDITION

FIRST PRINTING

Printed in the United States of America

Library of Congress Cataloging in Publication Data

Heiserman, David L., 1940-
Build your own working robot.

Includes index.
1. Robotics. I. Title
TJ211.H35 1987 629.8'92 86-30127
ISBN 0-8306-1181-9
ISBN 0-8306-2781-2 (pbk.)

Questions regarding the content of this book should be addressed to:

Reader Inquiry Branch
Editorial Department
TAB BOOKS Inc.
P.O. Box 40
Blue Ridge Summit, PA 17214

Contents

Preface

The world has changed a great deal since the publication of *Build Your Own Working Robot* (TAB book No. 841) in 1976. The world in general is a different place, and so is the world of robotics. A backward glance at the robotics of 1976 can be an exercise that is both instructive and enlightening. In any event, such an assessment can go a long way toward justifying the differences between this book and the 1976 book.

In 1976, there weren't many books on the subject of robotics, and only two described exactly how to build a working robot in a home workshop. Most books dealing exclusively with robotics were written on a professional level and generally amounted to little more than descriptions of how the then-popular "numerically controlled" machines could be made more versatile. That has all changed. Engineering colleges and technical schools still claim there aren't enough good textbooks on professional robotics, but amateur roboticists now have the luxury of choosing among a wide variety of practical books written especially for them.

Before 1976, very few people had any real enthusiasm for robots—fictional or otherwise. But a single event changed that situation, perhaps forever. Shortly after the publication of *Build Your Own Working Robot,* the film, *Star Wars,* began breaking box-office records and clearly defining the popular image of robots. The two main robotic characters, R2D2 and C3PO, as well as a host of fascinating mechanical walk-ons and extras, forged a popular image of robots that promises to set the standard for many years to come.

One of the spin-offs from *Star Wars*, its sequels, and imitations was a surge of interest in popularized robot machines, or "personal robots." The advent of the personal computer and video games had primed the pump of electronic gadgetry, and the time was ripe for a frenzied effort to produce and consume books, magazines, hardware, and computer software that bore any relationship to robots.

In the preface to *Build Your Own Working Robot*, I stated that the popular perception of a real robot expected too little, rather than too much, of the prevailing electronics technology. That is no longer the case. The new vision of robots set our sights far beyond the capabilities of most industries and certainly the basement experimenter. The enthusiasm for personal robotics crested in 1983 and plunged rapidly downward as the technological and economic realities of the matter became apparent. Personal robots simply cost too much for what little they could do.

1984 marked the beginning of a shake-out in personal robotics. New businesses devoted exclusively to the production of personal robots began losing a lot of money. Some have survived on the grounds that they represent educational tools. Some models are surviving in more modest forms, disguising the product as cute furry animals that bear no resemblance to the popular image of robots.

The commercial picture of personal robotics is certainly a depressing one. But through all of this, there remains a devoted core of amateur enthusiasts to carry the technology forward to its next heyday in the future. Such people aren't concerned about making a robot that fetches the morning paper, washes windows, sweeps the floor, and changes the baby's diapers.

In 1976, it was a new idea to suggest that a meaningful personal robot might be one that simply serves the purpose of a puppy dog—a companion to be played with and watched. Now that is an old and well-established notion that compells amateur experimenters to continue developing new machines and concepts. And nowadays, the "star wars" generation of experimenters finds it easy to develop a certain kind of affection for a machine that they have built from scratch. As a result, it is simpler these days to overlook the faults and shortcomings of one's own robot and point out the positive features (whether real or imaginary) with some justifiable amount of pride.

****

There aren't many books on the subject of robotics, and there are even fewer books telling exactly how to build one in a home workshop. This book is about an unusual robot named Buster, and this little machine has been specifically designed for amateur experimenters who want to try their hand in an infant technology known as *robotics*.

Buster is a true robot. He is not a toy, and he is not one of the fancy remote-controlled props that have been used in popular science-fiction TV shows and movies over the past two decades. He is much more than a toy or a prop. Buster is more like an animal than a machine—he has some basic reflex mechanisms, a will of his own, and even a personality of sorts. He does not merely mimic animal behavior, either; rather, he is a creature in his own right, and he acts and reacts according to a dynamic programming mechanism that is sensitive to his internal and outside environments.

It is rather difficult to describe exactly what a robot is. The main problem is that there are so many prevailing misconceptions and preconceived notions about what a robot ought to be, and unfortunately, these popular ideas generally represent too modest, rather than too advanced, a concept of these machines. Because there is a need to change many popular idea, some of the material in this book reads more like philosophy and psychology than electronics. To be sure, the experimenter will find out how to build Buster for himself, but he will miss an appreciation of what he is really doing if these nontechnical discussions are overlooked or treated lightly.

The experimenter planning to build the Buster system ought to have a background in basic electronics, including a familiarity with TTL technology, transistor amplifiers, and elementary control circuits. A working knowledge of Boolean algebra is also important for understanding and troubleshooting some of the more complex circuits. It is equally important to have mastered some of the basic hands-on skills of electronics, such as making printed-circuit boards.

Building Buster is an ambitious undertaking. There is, however, no need to build the entire system before putting it to work. The book is so planned that an experimenter can build up the system one step at a time, with each step adding a new dimension to Buster's behavior patterns. An experimenter, for instance, might complete only half the work, but even so, he will have a fascinating machine that runs around the floor, pokes into corners, and cries woefully whenever his batteries run low. An experimenter, in other words, can do as much of the work as time and finances allow, and if it takes five years to finish the job, so what? He's had a lot of fun working, playing, and observing an example of the highest class of machine technology can produce today.

I am indebted to a number of individuals who helped with the construction of the prototypes and preparation of the manuscript. A special word of appreciation must go to the Ohio Institute of Technology in Columbus and the many staff members and students who made it possible to engineer, build, and debug the Buster system in such a short time.

Chapter 1

The Buster Concept

The purpose of this chapter to help you become familiar with what you can expect of a finished Buster system and suggest some approaches to constructing such a system from scratch. The idea is to prepare you for an ambitious and time-consuming undertaking, and help you decide whether or not you think the rewards are worth the effort.

GENERAL DESCRIPTION OF THE BUSTER SYSTEM

The Buster system described in this book is not very big. He is built around a circular platform that is about 13 inches in diameter, and he is about 14 inches tall. Your own version might be slightly larger, depending on the selection of some of the major components. (Incidentally, the fact that the diameter of the platform depends on the selection of major components strongly suggests that you gather all the main parts—motors, wheels, and batteries—before fixing the dimensions of the platform for your own project.)

So Buster isn't very big in a physical sense. But he is a big project that can mean a lot to you, both now and in the future. The following discussions outline some of his main technical features.

The Ways Buster Moves

Buster expresses his personality in the ways he moves around on the floor. His physical responses to commands from his human r and automatic responses to conditions in his immediate onment all take the form of some sort of motion.

The Buster platform is supported by four wheels: two motor-driven wheels and two idler wheels. The motor-driven, or drive wheels are located on the left and right sides of the platform. The idler wheels are located at the front and rear.

The drive wheels are operated independently of one another. For example, the left motor can be run in a forward direction while the right motor is stopped. Using a pair of independently driven motors in that configuration allows them to handle both the motion and steering operations.

Each drive motor can be run in a forward direction, a reverse direction, and stopped. Given a pair of such motors, you can see that Buster's platform can move the nine different ways shown in Table 1-1.

Obviously Buster is motionless whenever both drive motors are stopped. He moves straight forward when both motors are running forward, and he moves straight backward when both are running in reverse.

The platform exhibits two different kinds of turning motions. One kind of turn is about the center of the platform, and the other is about one drive wheel or the other.

Turning the platform about its center is a matter of running the motors in opposite directions. Running the left motor in its forward direction and the right motor in reverse, for example, spins Buster in the clockwise direction about the center of the platform. On the other hand, running the left motor in reverse and the right motor forward causes Buster to spin about the center of the plat-

Table 1-1. Summary of Buster's Platform Motions.

Left Motor	Right Motor	Platform Response
STOP	STOP	FULL STOP
STOP	FORWARD	FORWARD WITH LEFT TURN
STOP	REVERSE	REVERSE WITH LEFT TURN
FORWARD	STOP	FORWARD WITH RIGHT TURN
FORWARD	FORWARD	STRAIGHT FORWARD
FORWARD	REVERSE	CLOCKWISE SPIN
REVERSE	STOP	REVERSE WITH RIGHT TURN
REVERSE	FORWARD	COUNTER-CLOCKWISE SPIN
REVERSE	REVERSE	STRAIGHT REVERSE

form in the counter-clockwise direction.

The second kind of turning is the result of running one motor while stopping the other. The turn axis is about the motor that is stopped. Consider running the left motor in the forward direction while stopping the right motor. In that case, the platform moves forward, but in an arc having the right motor as its center.

Blunder Buster

Buster's motion can be entirely determined by commands from a human operator or entirely dictated by contacts with objects in his immediate environment. A third option is to configure Buster so that he responds to commands until he runs into something and, at that time, override the commands so that he can handle the situation in his own fashion.

Unless Buster is configured so that he responds entirely to commands from the human operation, he must possess some means for sensing and evaluating physical conditions in his immediate environment. That is the purpose of Buster's contact-sensing scheme.

Buster's contact-sensing scheme consists of a set of four lightweight bumpers that operate sets of microswitches. Depressing one of the bumpers closes the corresponding microswitch.

The scheme uses just four blunder bumpers—front, right side, rear, and left side. Whenever Buster blunders into an obstacle such as a wall or a piece of furniture, one or more bumper assemblies force their switches closed. The fact that any blunder switch is closed indicates some sort of blunder condition, and the pattern of switch closures indicates the nature of the contact.

Table 1-2 summarizes all 16 possible combinations of blunder conditions.

The Blunder-Response Programming

One of the main features of the new Buster system is that he is not software programmable. That is, he does not require any sort of computer programming in order to execute his full range of activities.

However, Buster must possess some sort of internal programming that links the blunder system with his motor system. He hardly qualifies as a robot without some minimal sort of built-in "smarts."

The system gets around the need for computer programming and yet exhibits "smart" responses to blunder conditions by means of some hardwired programming—programming that is perma-

Table 1-2. Summary of All Possible Blunder Conditions.

Bumper(s) Activated				Blunder Pattern
Front	Right	Left	Rear	
*				none
	*			FRONT only
*	*			RIGHT only
				FRONT and RIGHT
		*		LEFT only
*		*		FRONT and LEFT
	*	*		RIGHT and LEFT
*	*	*		FRONT, RIGHT and LEFT
			*	REAR only
*			*	FRONT and REAR
	*		*	RIGHT and REAR
*	*		*	FRONT, RIGHT and REAR
		*	*	LEFT and REAR
*		*	*	FRONT, LEFT and REAR
	*	*	*	RIGHT, LEFT and REAR
*	*	*	*	FRONT, RIGHT, LEFT and REAR

nently fixed by the patterns of wiring between electronic logic and control circuits. Table 1-3 summarizes Buster's permanently programmed responses to all possible combinations of blunder conditions.

Operating Modes

The operator has a choice of four different operating modes for the Buster system. Each operating mode determines the general personality of the machine—the way Buster reacts to instructions from a human operator, responds to contacts with obstacles in the environment, and generally conducts himself.

The human operator selects an operating mode by setting the positions of a pair of mode-select toggle switches that are mounted on the Mode Switch and Lamp Panel. The following paragraphs describe the general nature of the four operating modes.

Mode 0 is a standby mode. Buster sits motionless until some external event activates a blunder-bumper switch assembly. When that sort of event occurs, Buster executes the automatic blunder sequence prescribed for that particular contact situation. And he continues reacting in that way until none of the bumper switches is closed.

In a manner of speaking, Mode 0 puts Buster to sleep. Power

consumption is at a minimum, and from all outward appearances, he seems to be turned off. But he is primed and ready to respond instantly to any contact with the blunder switches.

The 1976 version of Buster used a similar operating mode called the *full-function standby mode*. The family cat once approached Buster in that mode. The cat's cautious curiosity suddenly turned to terror, however, as he touched one of the bumpers and found himself being chased by a growling mechanical monster.

Nearly a decade later, that same orange tomcat is still a quiet companion in my workshop. But his experiences with a succession of different robot machines through the years have instilled a general mistrust for such machines, and he has never approached the new Buster model.

Mode 1 gives the human operator full control over Buster's motion. The blunder feature is disabled, so it is up to the operator to maneuver Buster through contact situations in the environment. The operator sends motion-control instructions to the machine by means of a hand-held switch panel and a long umbilical cord.

Mode 1 was called the *complete local command mode* in the 1976 version. That mode, then and now, generally violates some fundamental philosophical definitions of robotics by overriding all

Table 1-3. Summary of Preprogrammed Blunder Responses to All Possible Blunder Conditions.

Blunder Condition	Programmed Buster Response
none	STRAIGHT FORWARD
FRONT	REVERSE WITH RIGHT TURN
RIGHT	FORWARD WITH LEFT TURN
FRONT/RIGHT	REVERSE WITH LEFT TURN
LEFT	FORWARD WITH RIGHT TURN
FRONT/LEFT	REVERSE RIGHT
RIGHT/LEFT	STRAIGHT FORWARD
FRONT/RIGHT/LEFT	STRAIGHT REVERSE
REAR	STRAIGHT FORWARD
FRONT/REAR	CLOCKWISE SPIN
RIGHT/REAR	FORWARD WITH LEFT TURN
FRONT/RIGHT/REAR	COUNTERCLOCKWISE SPIN
LEFT/REAR	FORWARD WITH RIGHT TURN
FRONT/LEFT/REAR	CLOCKWISE SPIN
RIGHT/LEFT/REAR	STRAIGHT FORWARD
FRONT/RIGHT/LEFT/REAR	STOP

"smarts" that are built into the machine. It is nevertheless a convenient operating mode to use whenever it is important to move Buster from one place to another without giving him an opportunity to snoop around among obstacles that might fall into his path.

Operating Mode 2 represents something of a compromise between a machine that is fully under the control of a human operator and one that is able to exercise some "smart" reflex behavior. The mode was described as *local command with reflex priority* in the earlier version of the Buster system.

Mode 2 lets the human operator control the machine—via a control panel and umbilical cord—but only as long as Buster is not called upon to deal with a contact situation. The operator can give Buster some general direction of motion, but the machine ignores those instructions while dealing with obstacles in his path.

Mode 3 sets Buster loose on his own. His activity becomes continuous. He generally runs straight forward, but his path is changed by the nature of his automatic response to contacts with walls, furniture and any other obstacle that can activate the blunder-bumper switches.

The free-running, Mode-3 behavior gives Buster an element of independence that is a rare characteristic in modern-day machines. In fact, you do not have to devise ways to make Buster do interesting things—the problem is keeping him from overextending himself with activity. Buster behaves very much like an excited puppy: fun, overactive and unpredictable. The only way to quiet him down is to chase after him and reset the mode switches to a different operating mode (or let the batteries run down).

HOW TO APPROACH THIS PROJECT

If, by the time you reach this point in your reading, you are still interested in pursuing the Buster project, the next step is to study Chapter 2 very carefully. That chapter describes the physical configuration of the main platform and the nature of some critical components—components that determine how you will have to go about starting the construction phase of the job.

The problem you must solve first is locating a pair of suitable drive wheels. Unfortunately there is no reliable source of standardized motors and wheels for personal robot projects. Discussions in Chapter 2 describe those motors in a general sense, but you must seek your own source. And you must have them at hand before laying out the basic platform.

Another essential component, the batteries, are far easier to find and are available in reasonably standardized forms. If you use a pair of motorcycle batteries, as suggested in Chapter 2, they will possess enough electrical punch to do the job and yet be small enough to fit onto the suggested platform.

In all respects, the Buster machine is actually built around the gearmotors, wheels and batteries. Get those components into place, and there are few remaining questions regarding the size of your machine and the layout of the rest of the parts. That is why a thorough study of Chapter 2 is so important at the outset.

While you are in the process of rounding up the motors and batteries for your Buster project, I suggest you read the introductory discussions, circuit descriptions, and suggested construction hints in the remaining chapters. The idea serves a twofold purpose: to give you a broader picture of how the Buster system works and provide a clear idea of what to expect in terms of future construction tasks.

Chapter 3 deals with the Motor Output board. That part of the system represents the electronic interface between the digital circuits and the drive motors. Chapter 4 through 9 deal mainly with circuits that are located on the main electronics board, and in a couple of instances, additional panels and circuit boards. Note that Appendix A shows a suggested layout of components on the main electronics board. Chapter 10 marks the point in the project where you complete the project by linking the electronics to the motors.

If you are committed to the project during this phase of the work, you might want to begin gathering the small components as well. The specifications for the components are cited in the chapters that describe their application; but for the sake of ordering parts, you should use the complete parts list for the system in Appendix B.

When you have obtained suitable motors and batteries, and after you've had a chance to preview the work you will be doing, you are ready to begin putting things together and testing each step as you go along.

Work through the book one chapter at a time. Most of the chapters are organized in such a way that you are expected to understand exactly how a given part of the system works and what is involved in making it work. Then there is a list of suggested construction procedures followed by a set of suggested testing procedures.

The suggested testing procedures are especially important to the Buster project. They can be helpful long after you have completed the project, but they are an essential part of the construction procedure itself. The last-built part of the circuit should be thoroughly tested and put into good working order before going to the next part of the project. Failing to do that, you run the risk of having to troubleshoot a finished system that is so full of bugs that you can become frustrated trying to get it to work.

A WORD OR TWO ABOUT PERFECTION

Nothing is perfect; and, believe it or not, that applies to authors and the technical books they write. I have designed and built enough robots, and written and published enough books about them to know that there is no way in the world to avoid errors. Any author who takes the task seriously is fully aware of his or her responsibility to the readers. But mistakes do happen. There are no excuses; only simple facts of life.

Chances are good that there are going to be some errors in the presentations in this book. Some will be quite obvious and easy to circumvent—citing R101 in place of C101, for example, or making a typographical mistake such as using the word "now" instead of "not."

Other kinds of slip-ups can be far more confusing, however. Suppose that a suggested testing procedure says that you are supposed to find a logic-1 level at a certain test point, but the test should actually show a logic-0 level. Your circuit might be working perfectly, but an error in the book suggests otherwise.

Hopefully there will be no errors of that kind, but the presentations have some built-in safeguards. Just about every little element of the Buster system is presented in several different ways: parts lists, schematics, circuit descriptions, test and construction procedures, and even summaries of how the components are supposed to work. If you suspect you have found an error, you should take a close look at how that point in the circuit is described elsewhere and in other ways. Making a conscientious effort to resolve the matter, you will either confirm your suspicion, discover that you had misunderstood some of the theory or made an error of your own.

Chapter 2

Constructing the Moveable Platform

This chapter deals with the major portion of all the mechanical work required for putting together the Buster system: constructing the main platform and power distribution board; mounting the drive/steer wheels, idler wheels, batteries, and blunder bumpers; and connecting terminal strips and wiring for the motors, batteries, and blunder switches.

Table 2-1 lists the electrical components that are directly associated with the platform assembly. See Table 2-2 for components that are to be mounted on the power distribution circuit board.

The fact that there is such a great diversity of shapes and sizes for the motors, batteries and blunder switches makes it difficult to specify exact dimensions and mounting techniques. So the instructions offered here should be regarded as suggestions rather than exact construction details.

CONSTRUCTING THE MAIN PLATFORM

The drive/steer scheme suggested for this Buster project requires the use of a circular main platform. And given the shapes and dimensions of readily available batteries and drive/steer motors, the main platform ought to be approximately 13 inches in diameter.

When selecting a material for the main platform, bear in mind that it should be both sturdy and lightweight. An ideal material is

Table 2-1. Parts List for Platform Electrical Components.

M1, M2—6 Vdc gearmotor wheel assembly
B1, B2—6 V motorcycle battery
S2, S3, S4, S5, S6, S7, S8, S9—Lever-type microswitch (such as Radio Shack #275-017)
TS1—4-position, dual-row barrier strip (such as Radio Shack #274-658)
TS2—6-position, dual-row barrier strip (such as Radio Shack #274-659)

1/16-inch sheet aluminum stock; it is indeed sturdy and lightweight, but it is also relatively expensive and sometimes hard to find.

An alternative to aluminum stock is the 1/4-inch fiberboard that is commonly used as the base material for wall paneling. This material is fairly sturdy and lightweight and is inexpensive and easy to form.

Standard plywood stock is certainly sturdy, inexpensive, and is easy to find and form into the desired circular shape. But on the other hand, it is relatively heavy. If you find you are forced to work with plywood, I suggest you use the thinnest you can find.

Regardless of the material you choose for constructing the main platform, the procedure for laying out the circle is the same. The general idea is to layout the circle on a sheet of stiff cardboard—posterboard, for instance. Carefully cut and trim the circle from the cardboard, and then use it as a template for cutting the platform material.

Lay the uncut cardboard onto a flat, smooth surface. Tie a piece of string around the shaft of a common thumbtack and stick the tack through the cardboard at the desired center point of the circle. Measure off 6 1/2-inches of string from the thumbtack and hold that point firmly against the tip of an ordinary pencil. With a bit of practice, you can scribe a decent circle on the cardboard.

When you're satisfied with the quality of the circle, cut it out with a pair of scissors, and if necessary, patch and trim any rough places.

Table 2-2. Parts List for Power Distribution Circuit Board.

S1—DPST toggle switch; contact rating is 3 A minimum
J1—Red binding post (such as Radio Shack #274-662)
J2—Black binding post (such as Radio Shack #274-662)
F1, F2—5 A slow-blow fuse
TS3—6-position, dual-row barrier strip (such as Radio Shack #274-659)
2 Fuseholders

Lay the cardboard template onto the material you are using for the platform. Find the center of the circle (the hole left by the thumbtack) and punch a mark through the template and into the platform material. Then score the outline of the circle. If you are using aluminum, you should scribe the circle with a sharp, pointed instrument. Otherwise a pencil will do the job.

Sawing the circle from wood or fiberboard is a straightforward procedure. If you aren't very good at sawing such materials, make a rough cut first, then use a file or sandpaper to refine the shape.

It can be difficult to cut a circular shape from aluminum stock if you are using a hacksaw. Ideally, you can cut the circle with a sabersaw that is outfitted with a suitable, metal-cutting blade. A nibbling tool works rather well, too. In any event, you will probably end up having to knock off some rough corners with a file.

Finish the edges with sandpaper or crocus cloth. If you are using wood or fiberboard, it is a good idea to finish both surfaces and the edge with several coats of a tough acrylic spray.

MOUNTING THE GEARMOTORS AND IDLER WHEELS

Proper platform motion is possible only when the two drive/steer wheels are located exactly on opposite ends of a straight line that runs through the center of the circle. The idler wheels should also be located on a straight line through the center and at a 90° angle from the line used for the drive/steer wheels.

Locate the mark that indicates the center of the circle, and use a straightedge to score a diameter from edge to edge and through the center point. Score an identical line—edge-to-edge and through the center—but at right angles to the first. A carpenter's square can be helpful in that regard.

With the platform laying flat on a smooth surface, position the drive/steer wheels at opposite ends of one of the diameters you have just drawn. The exact procedure for mounting the drive/steer wheels and gearmotors depends on their actual mechanical configuration, so you have to figure out the best way to lay out and cut the mounting holes.

A few general hints are in order, however. First, the axles of the wheels, themselves, must be located directly over the diameter line on the platform and aligned parallel to it. Second, the drive/steer wheel assembly should be arranged so that the wheels are closer to the outside of the platform assembly. And finally, the wheels should be set in about 1 inch from the outer edge of the platform. (See Fig. 2-1.)

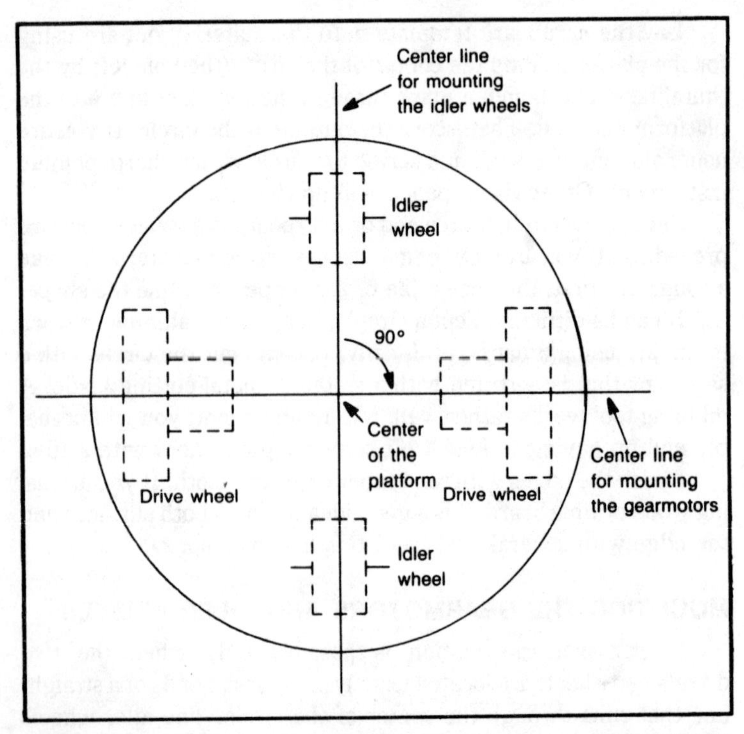

Fig. 2-1. Layout of the platform for mounting the drive/steer and idler wheels.

These hints are based on the fact that the axle of the wheel, and not the shaft of the motor or gearing mechanism marks the point of actual motion. Furthermore, the motor is generally fixed on one side of the wheel, and the only way to bring the wheel near the edge of the platform (for mechanical stability) is to mount the assembly so that the motor faces the center of the platform.

Unless common sense for your own situation dictates otherwise, the drive/steer wheels should be temporarily mounted to the platform before laying out the mounting scheme for the idler wheels.

Position the idler wheels at opposite ends of the remaining diameter line. Make sure the idling axes are positioned directly over that line and that the wheels do not rotate any more than 1 inch from the edge of the platform. Temporarily mount them to the platform.

The final step is to make sure that the platform is stable and horizontal when resting on the four wheels. Chances are good that you will have to shim one or the other set of wheels. If the idler wheels are too small, for example, fit some wooden or aluminum spacers between their mounting brackets and the bottom of the platform. On the other hand, you might find it necessary to use some spacers between the drive/steer wheel assemblies and the platform in order to lower their axles. Simple common sense and some patience will help you work out any wobbling effect.

Once the drive/steer and idler wheels are permanently mounted, drill holes through the platform and feed the wiring from the motors through them. Use vinyl feed-thru grommets if the platform is made of metal.

MOUNTING AND TESTING THE BLUNDER BUMPERS

The Buster system calls for mounting four blunder bumpers that are activated by mechanical pressure at the front, rear and sides of the platform. The bumpers themselves, are sections of 1/8-inch aluminum wire—clothesline wire. Each bumper uses two contact switches; so the objective is to mount four bumpers and four pairs of lever-type microswitches at 90° intervals around the perimeter of the platform.

The blunder mechanism requires some custom mounting brackets. You will need a right-angle mounting bracket for each of the eight microswitches and four other brackets that serve as connecting points for the bumpers. Figure 2-2 suggests shapes and mounting arrangements for the items that make up the blunder-bumper scheme. The switches and mountings are located on the bottom surface of the platform—assuming of course that your wheels are large enough to allow an inch or two of clearance from the floor.

Figure 2-3 shows the general, top-view layout of the platform as it should appear at this point in the construction procedure. Notice that the bumper connectors are located 90 degrees from one another, and 45 degrees from the center lines that were used for mounting the drive/steer and idler wheels.

Position and mount the bumper connectors first. Then mount a microswitch on both sides of each bumper connector, making sure that the switch lever is pointing away from the bottom surface of the platform. Cut and form the four bumper elements, feed the ends through the appropriate holes in the bumper connectors and bend the ends, so that the bumpers are loose but do not fall out of the connectors.

When this phase of the job is done, you should be able to close the bumper switches by pressing on the bumper elements.

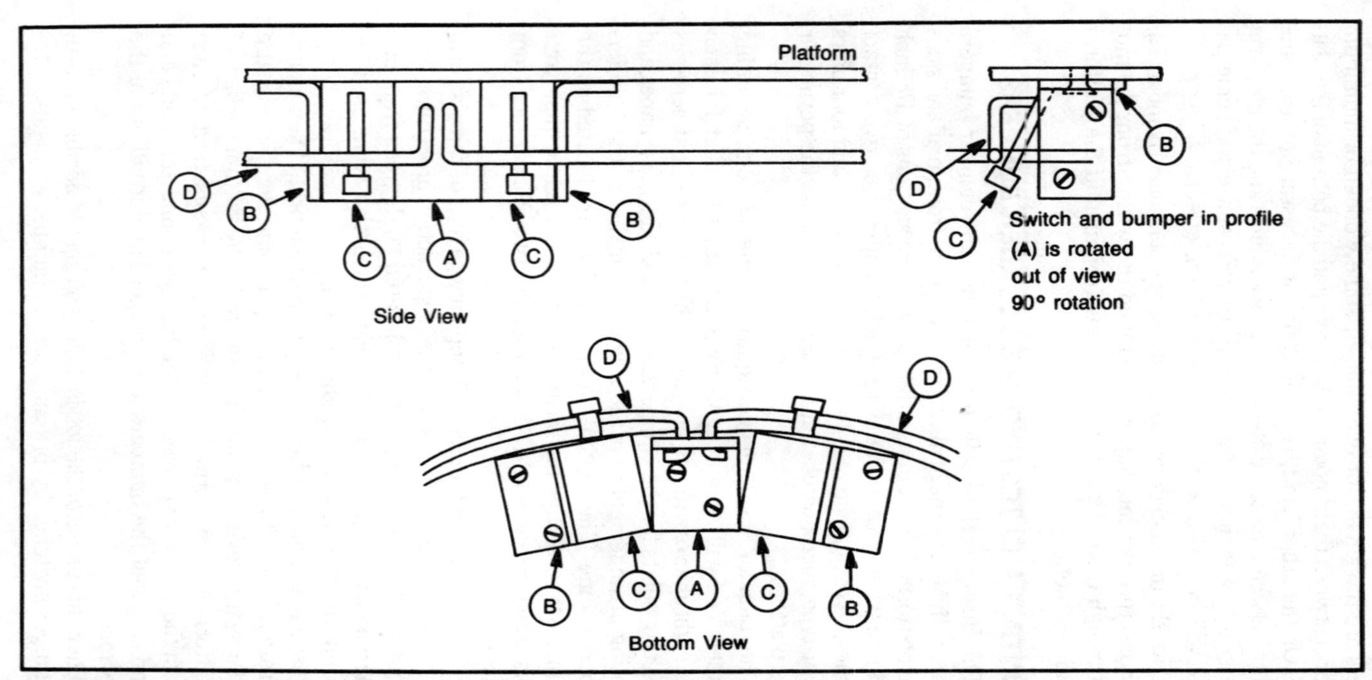

Fig. 2-2. Side view, bottom and rotated views of blunder-bumper components. Individual items are identified as follows: (A) Custom bumper connector; (B) Microswitch mounting bracket; (C) Lever-type microswitch; (D) Bumper element.

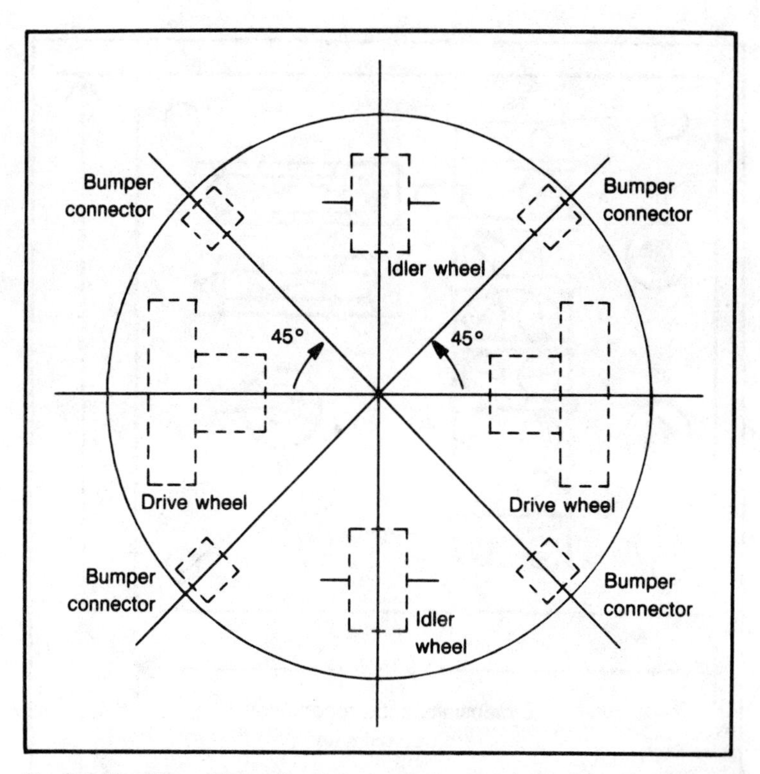

Fig. 2-3. Top view of the main platform showing locations of the wheels, and bumper connectors.

CONSTRUCTING AND MOUNTING THE POWER DISTRIBUTION BOARD

Table 2-2, Fig. 2-4, and Fig. 2-5 completely represent the Power Distribution Board: the table specifies the parts, and Fig. 2-4 suggests their layout on the board. Figure 2-6 is a schematic diagram of the wiring.

The binding posts are the connecting points to the external battery charger. The positive terminal of the battery charger connects to the red post, J1, and the negative terminal goes to the black one, J2.

Switch S1 is Buster's main power switch. It is a double-pole switch that is wired in such a way that one pole switches the +6 V tap from the batteries and the other switches the +12 V tap.

Fuses F1 and F2 protect the system against potential short-circuit conditions in circuitry that is yet to be installed. F1 protects the +12 V electronic circuits, and F2 protects the +6 V motors

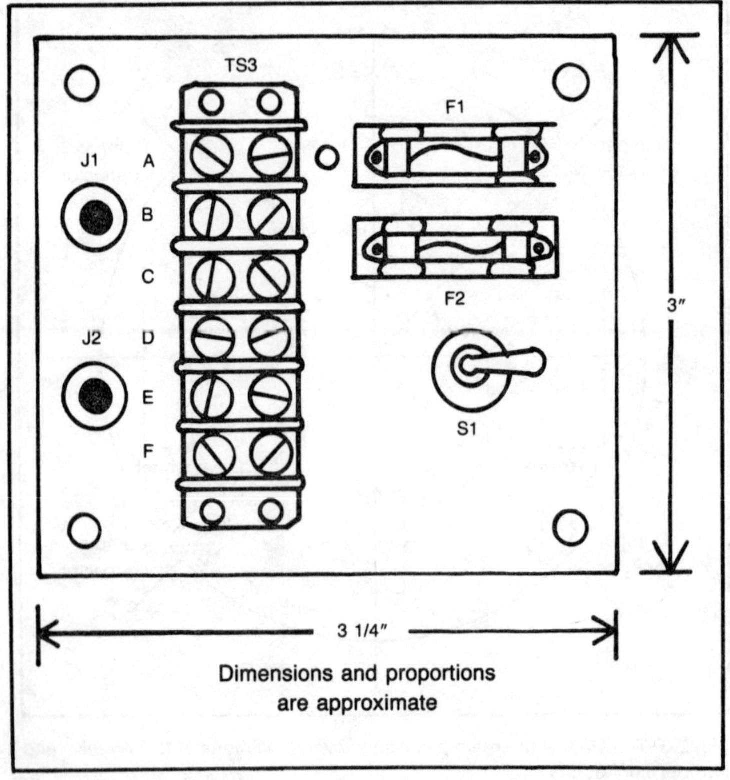

Fig. 2-4. Suggested layout of components on the Power Distribution Board.

and their associated circuitry.

The board can be cut from aluminum stock, 1/4-inch fiberboard, or even PC board stock. The dimensions shown on the mechanical layout are approximate, and you should make sure your own selection of components fit onto the board before cutting and drilling it.

Mount the components onto the board, then use the schematic in Fig. 2-5 as a guide for making solder connections to the components and solderless connections to the barrier strip.

At this point in the construction procedure, you have not really defined Buster's front and rear edges, and his left and right sides. Placement of the Power Distribution Board finally defines matters. The board should be located near the rear edge as indicated in Fig. 2-6. Having done that, you clearly define the front and rear edges and the left and right drive/steer wheels.

The completed Power Distribution Board should be mounted securely to the top surface of the platform. Use some sort of spacer

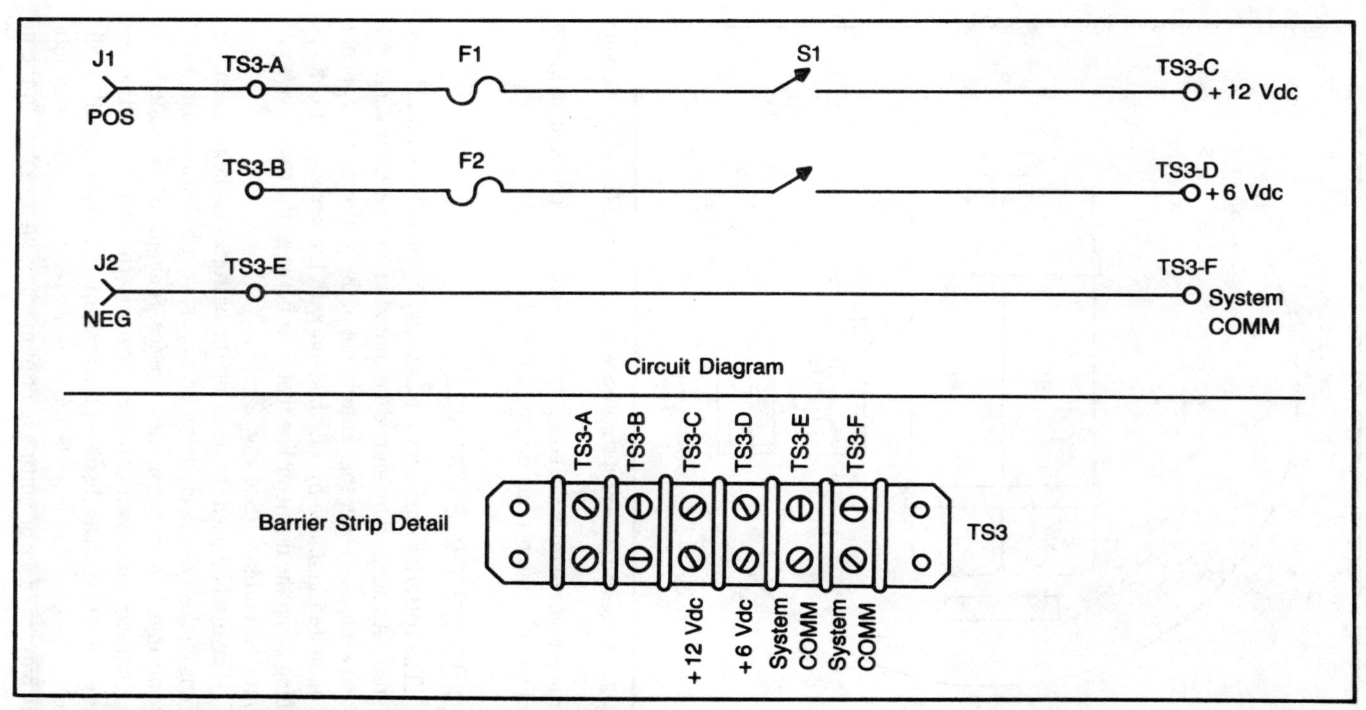

Fig. 2-5. Schematic diagram for the Power Distribution Board.

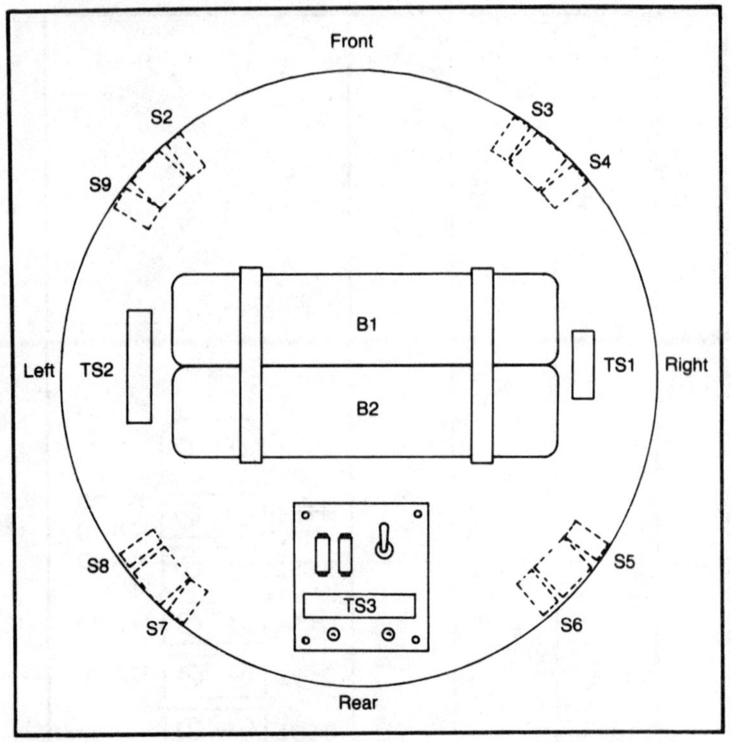

Fig. 2-6. Suggested placement of components on the top surface of the platform.

arrangement to raise the board in such a way that the binding posts and switch connections do not touch the platform.

MOUNTING THE BATTERIES

The batteries represent the heaviest items in the entire Buster system. As such, they should be mounted so that their center of gravity is located over the center point of the platform. They should be mounted in a side-by-side fashion with the surfaces of contact falling along the diameter line that runs between the two drive/steer motor assemblies. (See Fig. 2-6.)

The method used for securing the batteries to the platform depends on the way the electrical terminals, vent holes, and filler holes are arranged on the top of the batteries. A single, home-made metal bracket can hold both batteries firmly to the platform. Other arrangements, such as flexible plastic straps, can work equally well. Just make sure that your design does not require drilling holes into the battery cases or putting a lot of mechanical pressure on them.

WIRING AND TESTING THE BLUNDER SWITCHES

The first step in this procedure is to mount the blunder switch terminal board, TS2, to the top of the platform and drill holes for feeding wiring from the blunder switches to that terminal. It can be helpful to remove the batteries from the platform during these steps:

- Mount TS2 flush with the top surface of the platform, just to the left of the batteries as shown in Fig. 2-6.
- Drill a 1/4-inch hole through the platform near each of the four bumper connectors, and if the platform is made of metal, insert a vinyl feed-through grommet into each of the holes.

Most microswitches have a SPDT format—a normal open (N/O) connection, a normally closed (N/C) connection, and a common connection. These connections are usually labeled on the switch housing. Use AWG #18 stranded hookup wire for all of the following connections. Use the schematic in Fig. 2-7 as a guide for wiring the blunder switches.

- Solder lengths of stranded wire between the common connections on all eight blunder-bumper switches.
- Run one end of a length of stranded wire down through the hole for blunder switches S7 and S8 and solder one end to the common connections between those two switches.
- Layout, cut and connect the other end of the wire prepared in the previous step to the COMM terminal on TS2.
- Solder a length of wire between the N/O connections of blunder switches S2 and S3.
- Run one end of a length of stranded wire down through the hole for switches S2 and S9, and solder one end to the N/O connection of switch S2.
- Layout, cut and connect the other end of the wire prepared in the previous step to the BS0 terminal on TS2.
- Solder a length of wire between the N/O connections of blunder switches S4 and S5.
- Run one end of a length of stranded wire down through the hole for switches S3 and S4, and solder one end to the N/O connection of switch S4.
- Layout, cut and connect the other end of the wire prepared in the previous step to the BS1 terminal on TS2.
- Solder a length of wire between the N/O connections of blun-

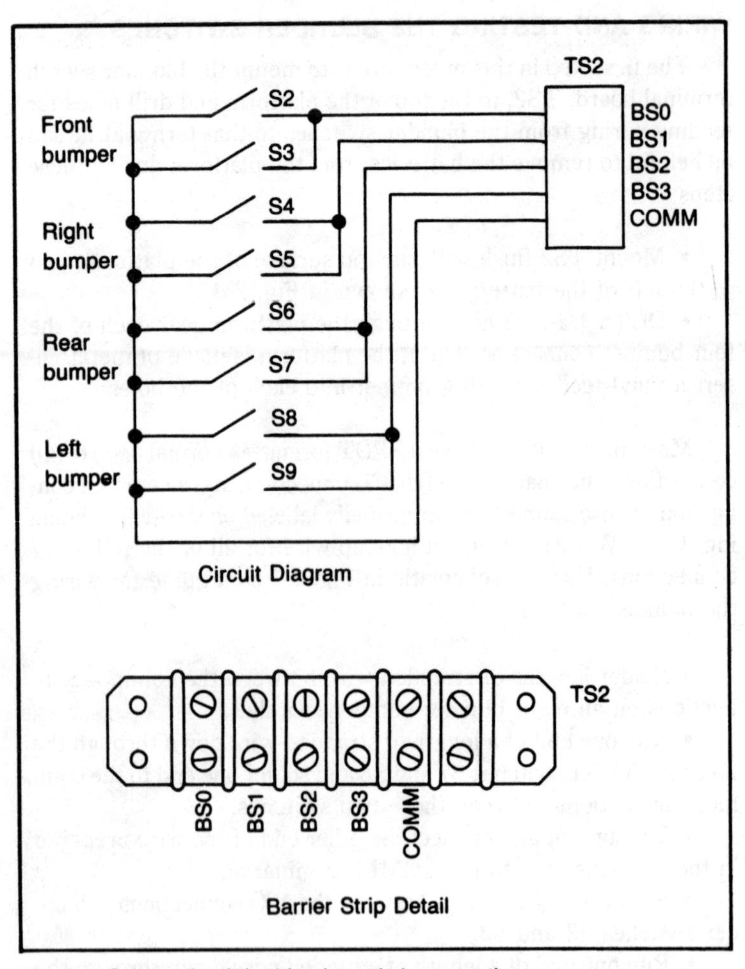

Fig. 2-7. Schematic of the blunder switch connections.

der switches S6 and S7.

• Run one end of a length of stranded wire down through the hole for switches S7 and S8, and solder one end to the N/O connection of switch S7.

• Layout, cut and connect the other end of the wire prepared in the previous step to the BS2 terminal on TS2.

• Solder a length of wire between the N/O connections of blunder switches S8 and S9.

• Run one end of a length of stranded wire down through the hole for switches S7 and S8, and solder one end to the N/O connection of switch S8.

• Layout, cut and connect the other end of the wire prepared in the previous step to the BS3 terminal on TS2.

An ohmmeter is the only instrument that is necessary for testing the blunder switch connections:

• Note the resistance between the COMM and BS0 terminals on TS2—zero resistance when you depress the FRONT bumper, and infinite resistance under any other condition.

• Note the resistance between the COMM and BS1 terminals on TS2—zero resistance when you depress the RIGHT bumper, and infinite resistance under any other condition.

• Note the resistance between the COMM and BS2 terminals on TS2—zero resistance when you depress the REAR bumper, and infinite resistance under any other condition.

• Note the resistance between the COMM and BS3 terminals on TS2—zero resistance when you depress the LEFT bumper, and infinite resistance under any other condition.

COMPLETING THE WIRING AND RUNNING FINAL TESTS

Complete the wiring for the drive/steer gearmotors as follows:

• Mount TS1 flush with the top surface of the platform, just to the right of the batteries as shown in Fig. 2-6.

• Drill a 1/4-inch hole through the platform near each of the two drive/steer gearmotors; and if the platform is made of metal, fit a vinyl feed-through grommet into each of the holes.

• If the gearmotors are not supplied with a capacitor already soldered to their electrical connections, solder capacitor C1 between the terminals on the left motor and C2 between the terminals on the right motor.

Figure 2-8 shows the schematic for connecting the gearmotors to terminal strip TS1. Use AWG #18 stranded wire for the following connections:

• Solder lengths of wire to each of the terminals on the left motor, M1.

• Feed the wires through the hole near motor M1, lay them out neatly, cut and connect them to terminals M1A and M1B and TS1. The order of connection is not important at this time—just

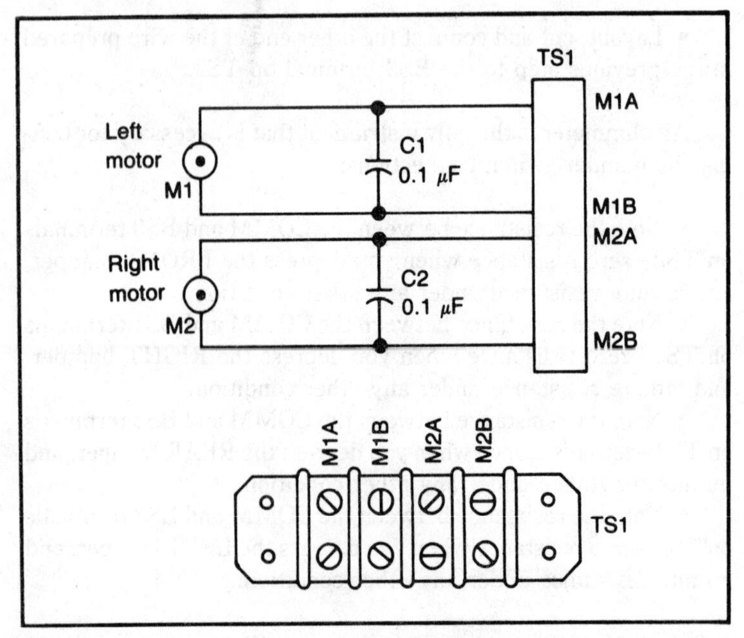

Fig. 2-8. Schematic of the connections between the gearmotors and terminal strip TS1.

make sure the wires are long enough to let you switch them if necessary.

• Solder lengths of wire to each of the terminals on the right motor, M2.

• Feed the wires through the hole near motor M2, lay them out neatly, cut and connect them to terminals M1A and M1B on TS1. The order of connection is not important at this time—just make sure the wires are long enough to let you switch them if necessary.

Secure the batteries into place and make the final connections between them and terminal strips TS2 and TS3. Use AWG #18 stranded wire and solderless connections in all cases:

• Connect the positive terminal of battery B1 to terminal TS3-A on TS3.

• Connect the negative terminal of battery B1 to terminal TS3-B on TS3.

• Connect the positive terminal of battery B2 to terminal TS3-B on TS3.

• Connect the negative terminal of battery B2 to terminal TS3-E on TS3.

• Connect a wire between terminals TS3-E and TS3-F on TS3.

• Connect a wire between terminal TS3-F on TS3 and the COMM terminal on TS2.

Test the integrity of the power connections with an ohmmeter:

• Note zero resistance between J1 and both terminals on fuse F1.

• Note zero resistance between J2 and COMM terminal on TS2.

• Note zero resistance between TS3-B on TS3 and both terminals on fuse F2.

• Note zero resistance between J1 and TS3-C on TS3 when switch S1 is ON, and infinite resistance when the switch is OFF.

• Note zero resistance between TS3-B and TS3-D when switch S1 is ON, and infinite resistance when the switch is OFF.

Connect the battery charger to binding posts J1 and J2—positive to J1 and negative to J2. Set up the charger for 12 V charging and turn it on. The meter on the charger should indicate that the batteries are taking a charge. Conduct the following voltmeter tests with the battery charger thus connected and in full operation:

• Measure 9 Vdc to 14 Vdc between binding posts J1 and J2—positive at J1.

• Measure 9 Vdc to 14 Vdc between terminals TS3-A and TS3-E on TS3—positive at TS3-A.

• Measure 4.5 Vdc to 7 Vdc between terminals TS3-A and TS3-B on TS3—positive at TS3-A.

• Measure 4.5 Vdc to 7 Vdc between terminals TS3-B and TS3-E on TS3—positive at TS3-B.

• Connect the negative lead of the voltmeter to TS3-F on TS3, and note 9 Vdc to 14 Vdc at TS3-C when S1 is closed and zero volts when it is off.

• Connect the negative lead of the voltmeter to TS3-F on TS3, and note 4.5 Vdc to 7 Vdc at TS3-D when S1 is closed and zero volts when it is off.

Chapter 3

Setting Up the Drive/Steer System

Buster expresses his personality in terms of the way he moves around on the floor. Of course the motion is produced by the two independent drive/steer wheels, so it follows that Buster's personality is the result of controlling these motors in meaningful ways.

Unfortunately the integrated circuits that set up the modes of motion do not produce nearly enough power to operate the motors directly. So it is necessary to include some power-boosting circuitry—specifically the Motor Output Board.

The work described in this chapter deals with the theory of operation, construction hints and testing procedures for Buster's Motor Output Board.

CIRCUIT DESCRIPTION

Figure 3-1 is a general schematic of a circuit that uses a pair of PNP transistors and a pair of SPDT relays to control the motion of a single motor. (The circuit actually used in the Buster system is shown in Fig. 3-2.)

Generally speaking, a relay is not energized as long as the transistor connected to its coil is switched off. The schematic shows the position of the relay contacts in that de-energized state—the normally closed (N/C) position. The relay is then energized when its transistor is switched to its conducting state, and the relay contacts move to the normally open (N/O) position.

The transistors are connected so that they are normally

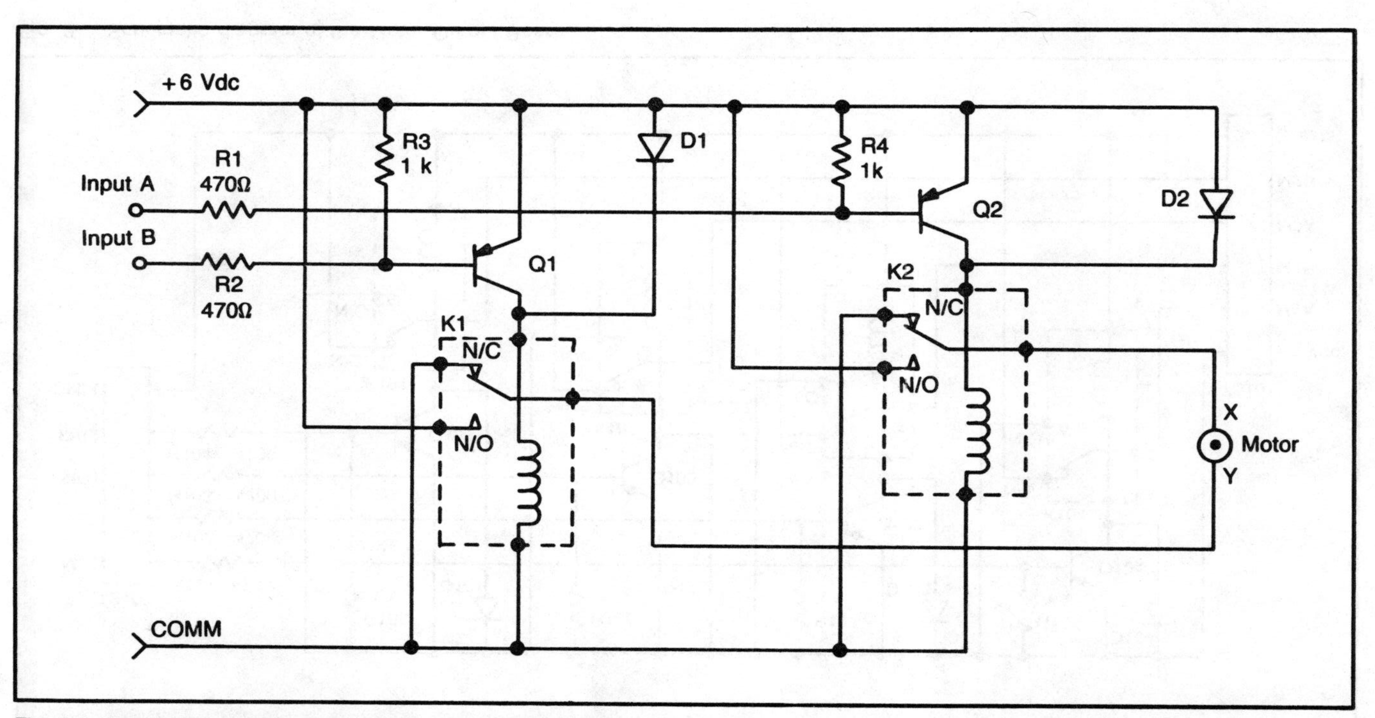

Fig. 3-1. General schematic of a simple motor control circuit.

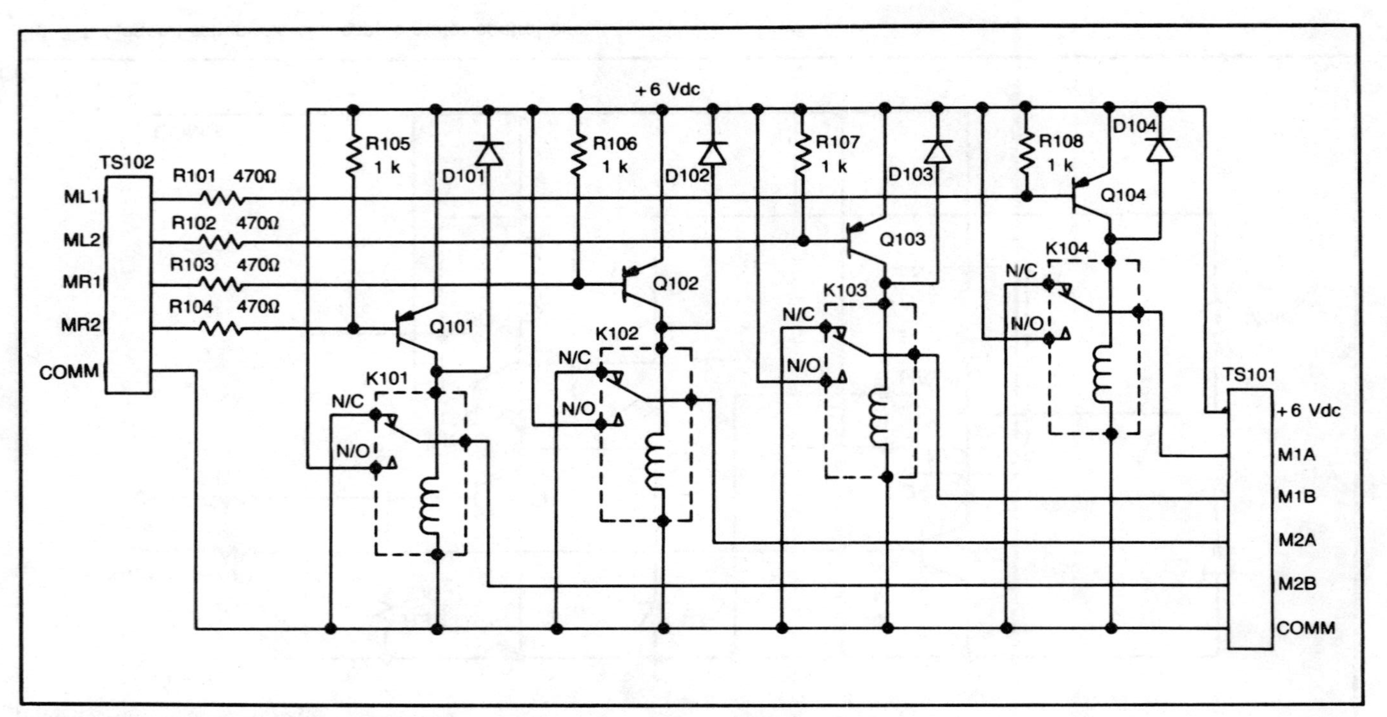

Fig. 3-2. Schematic diagram of the Motor Output Board.

switched off. That is the main purpose of the resistor connected between the base of each transistor and the +6 Vdc source. They are switched on by connecting their base connections, through a 470-ohm resistor, to the COMM line (the circuit's zero-volt reference).

Putting those facts together, a given relay contact is switched to its N/C position by grounding the input that is connected to the base of its transistor. The contact returns to the N/O position by making no connection to the input connection or connecting the base to the +6 Vdc source.

The moveable connection in the relays is connected to the motor. When both relays are de-energized, the moveable contacts in both relays are connected to the COMM line; and because that means there is no voltage difference between the terminals of the motor, it does not run. The motor does not run with both relays de-energized.

By the same token, the motor does not run when both relays are energized. Energizing both relays connects both motor connections to +6 Vdc—there is no difference in potential, and hence the motor cannot run.

The motor runs only when the two relays have opposite states: one is energized and the other is de-energized. Suppose that relay K1 is energized and K2 is de-energized. In this case, K1 supplies a +6 Vdc potential to contact Y on the motor, and K2 supplies a COMM (or 0V reference) potential to contact X. The motor thus has a 6 Vdc potential applied to it, and it responds by running in one particular direction.

Reverse the situation—de-energize K1 and energize K2—and the motor runs in the opposite direction. K1 applies the COMM potential to the Y terminal on the motor, and K2 applies +6 Vdc to the X terminal.

Whether the motor turns in reverse or forward is not important to the present discussion. The important point is that the motor runs when one input is connected to COMM and the other is not. The motor does not run when both inputs are connected to COMM, and it does not run when both inputs are connected to +6 Vdc, or nothing at all.

The purpose of the two diodes is to suppress reverse-voltage spikes that would otherwise appear across the emitter-collector terminals of the transistors whenever they were switched off. Such spikes are created by the collapsing magnetic fields generated by the coils in the relays.

Figure 3-2 shows the full schematic for Buster's Motor Output Board. Strictly speaking it is simply two sets of motor control circuits—one for the left motor and one for the right motor. The output of the circuit is at TS101, where connections M1A and M1B go to motor M1 (the left motor) and connections M2A and M2B go to motor M2 (the right motor).

The inputs to the circuit are connected to TS102. Input ML1 controls Q104 and relay K104, ML2 controls Q103 and relay K103, MR1 controls Q102 and K102, and MR2 controls Q101 and K101.

CONSTRUCTION HINTS

Table 3-1 specifies the components for the Motor Output Board, and Fig. 3-3 shows a suggested mechanical layout. The board, itself, is cut from printed-circuit perfboard. The small components can be glued directly to the board, with their electrical connections extending through the holes provided for that purpose.

Ultimately, the board is to be mounted vertically an inch or two in front of the batteries. This means you will also have to fashion a pair of suitable mounting brackets.

RUNNING PRELIMINARY BENCH TESTS

After you have mounted the components and made the necessary solder and solderless connections, it is a good idea to run some preliminary tests on the board before mounting it into place on the platform.

Use the following steps to set up the tests:

- Make sure there is a full charge on the batteries or that the battery charger is connected to the system and turned on.

Table 3-1. Parts List for the Motor Output Board.

R101, R102, R103, R104—470 Ω, 1/4 W resistor
R105, R106, R107, R108—1 kΩ, 1/4 W resistor
D101, D102, D103, D104—1N4001 diode
Q101, Q102, Q103, Q104—Any small, high-speed pnp transistor (such as 2N2907A)
K101, K102, K103, K014—SPDT relay; minimum 1 A contact rating; 6 Vdc coil having a current rating less than 100 mA (or dc resistance greater than 60 Ω)
TS101, TS102—6-position, double-row barrier strip (such as Radio Shack #274-670)

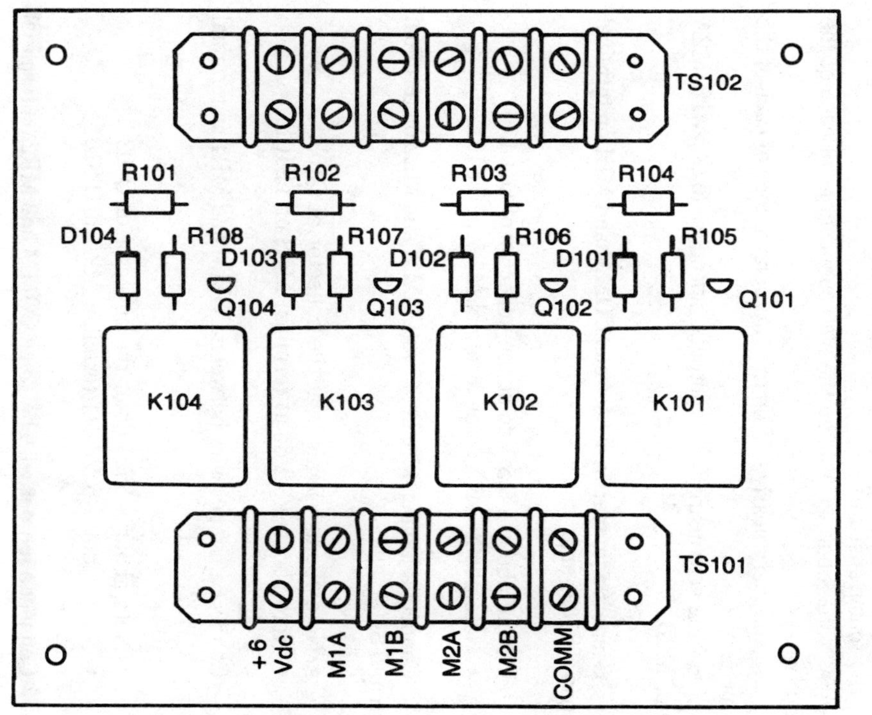

Fig. 3-3. Suggested mechanical layout of the Motor Output Board.

• Connect a temporary jumper wire between TS3-D on the Power Distribution Board and the +6 Vdc connection on TS101.

• Connect a temporary jumper wire between TS3-F on the Power Distribution Board and the COMM connection on TS101.

Make sure the main power switch is turned ON, and conduct the following voltage tests with the negative lead of the voltmeter connected to the COMM connection on TS101:

• Note about +6 Vdc at the +6 Vdc terminal on TS101.

• Note about +6 Vdc on the cathodes of each of the four diodes.

• Note about +6 Vdc on the emitter connections on each of the four transistors.

• Note very nearly 0 V on the collector connections on each of the four transistors.

• Note about +6 Vdc on the N/O connections of each of the four relays.

• Note very nearly 0 V on the N/C connections of each of the four relays.

• Note very nearly 0 V at output connections M1A, M1B, M2A and M2B.

Connect a jumper wire between COMM and ML1 on terminal strip TS102, and:

• Note about +5.5 Vdc at the collector of Q104.
• Note about +6 Vdc at terminal M1A of TS101.

Connect a jumper wire between COMM and ML2 on terminal strip TS102, and:

• Note about +5.5 Vdc at the collector of Q103.
• Note about +6 Vdc at terminal M1B of TS101.

Connect a jumper wire between COMM and MR1 on terminal strip TS102, and:

• Note about +5.5 Vdc at the collector of Q102.
• Note about +6 Vdc at terminal M2A of TS101.

Connect a jumper wire between COMM and MR2 on terminal

strip TS102, and:

- Note about +5.5 Vdc at the collector of Q101.
- Note about +6 Vdc at terminal M2B of TS101.

COMPLETING THE INSTALLATION

Mount the Motor Output Board vertically to the top surface of the platform and about an inch in front of the batteries. The board should be oriented so that the relays are closer to the platform and facing the front of the machine.

Use AWG #18 stranded hookup wire to make the following solderless connections:

- Layout, cut and connect a wire between TS3-D on the Power Distribution Board and the +6 Vdc terminal on TS101.
- Layout, cut and connect a wire between TS3-F on the Power Distribution Board and the COMM terminal on TS101.
- Connect a wire between the M1A terminal on TS101 and M1A on the motor terminal strip, TS1.
- Connect a wire between the M1B terminal on TS101 and M1B on the motor terminal strip, TS1.
- Connect a wire between the M2A terminal on TS101 and M2A on the motor terminal strip, TS1.
- Connect a wire between M2B terminal on TS101 and M2B on the motor terminal strip, TS1.

Conduct the following motor tests with a full charge on the batteries or when the battery charger is in full operation and connected to binding posts J1 and J2. Make sure the main power switch on the Power Distribution Board is turned ON, and prop Buster so that his drive/steer wheels are raised off the bench or floor.

- Connect a temporary jumper wire between COMM and ML1 on TS102, and note that the left motor runs.
- Connect a temporary jumper wire between COMM and ML2 on TS102, and note that the left motor runs.
- Connect a temporary jumper wire between COMM and MR1 on TS102, and note that the right motor runs.
- Connect a temporary jumper wire between COMM and MR2 on TS102, and note that the right motor runs.

There is no way of telling in advance which way the motors

Table 3-2. Summary of Motor and Platform Responses to All Possible Combinations of Input Logic Levels.

Input Level				Motor Response		Platform
ML1	ML2	MR1	MR2	Left	Right	Response
0	0	0	0	stop	stop	STOP
0	0	0	1	stop	forward	FWD LEFT
0	0	1	0	stop	reverse	REV LEFT
0	0	1	1	stop	stop	STOP
0	1	0	0	forward	stop	FWD RIGHT
0	1	0	1	forward	forward	FWD
0	1	1	0	forward	reverse	CW SPIN
0	1	1	1	forward	stop	FWD RIGHT
1	0	0	0	reverse	stop	REV RIGHT
1	0	0	1	reverse	forward	CCW SPIN
1	0	1	0	reverse	reverse	REV
1	0	1	1	reverse	stop	REV RIGHT
1	1	0	0	stop	stop	STOP
1	1	0	1	stop	forward	FWD LEFT
1	1	1	0	stop	reverse	REV LEFT
1	1	1	1	stop	stop	STOP

Inputs: 0 = logic 0 (connection to COMM)
1 = logic 1 (no connection or connection to +6 Vdc)

Platform Response:
STOP = full stop
FWD = straight forward
REV = straight reverse
FWD RIGHT = forward with right turn
FWD LEFT = forward with left turn
REV RIGHT = reverse with right turn
REV LEFT = reverse with left turn
CW SPIN = spin in the clockwise direction
CCW SPIN = spin in the counterclockwise direction

will turn when power is applied to them. The hookup has been arbitrary up to this point, but now it is time to define matters more precisely.

Table 3-2 summarizes Buster's full range of platform motions in terms of logic levels applied to the input terminals at TS102.

When you connect a jumper wire between the COMM and ML1 terminals on TS102, the left motor should turn in a direction that would carry Buster in a forward direction. If that is not the case, reverse the connections going to the left motor at M1A and M1B of TS1.

Likewise, when you connect a jumper wire between COMM and MR1 on TS102, the right motor should turn in a direction that would carry Buster in a forward direction. If that is not the case, reverse the connections going to the right motor at M2A and M2B of TS1.

Chapter 4

The Power Supply and Clock Circuits

The Power Supply and Clock circuits described in this chapter are located on the Main Electronics Board. Because this is the first series of construction steps that involve the Main Electronics Board, you should refer to the general notes in Appendix B.

Figure 4-1 is a schematic diagram of the Power Supply and Clock circuits. The circuits provide a regulated source of +5 Vdc for the entire circuit board and a set of clock pulses to be used by several other circuits in the system.

CIRCUIT DESCRIPTION

Voltage regulator VR200, transforms the +12 Vdc level from the Power Distribution Board to a regulated +5 Vdc source for all circuits included on the Main Electronics board. The regulator has a 1-amp capacity, making it more than adequate for the task.

The +12 Vdc source and COMM from the batteries are connected to the voltage regulator by way of barrier strip TS200. Capacitor C200 filters electrical noise from the motors. Such noise is normally filtered by the voltage regulator, itself; but even the regulator needs a bit of help in this regard when the batteries are operating below 60 percent of their full-charge level.

Capacitors C201 and C202 also help stabilize the +5 Vdc line, but against voltage fluctuations generated by the circuitry rather than low-battery conditions. C202, an electrolytic capacitor, filters longer-term voltage fluctuations. Electrolytic capacitors have a cer-

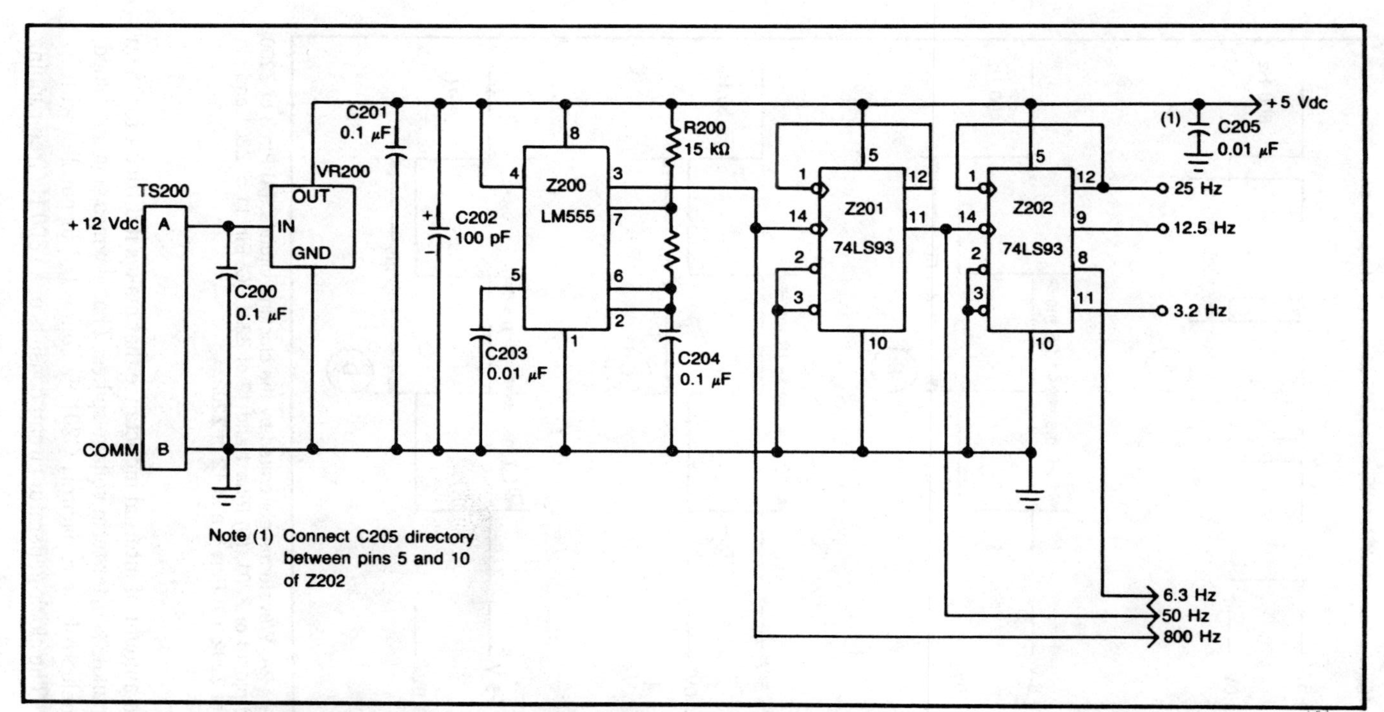

Fig. 4-1. Schematic of the Power Supply and Clock Circuits on the Main Electronics Board.

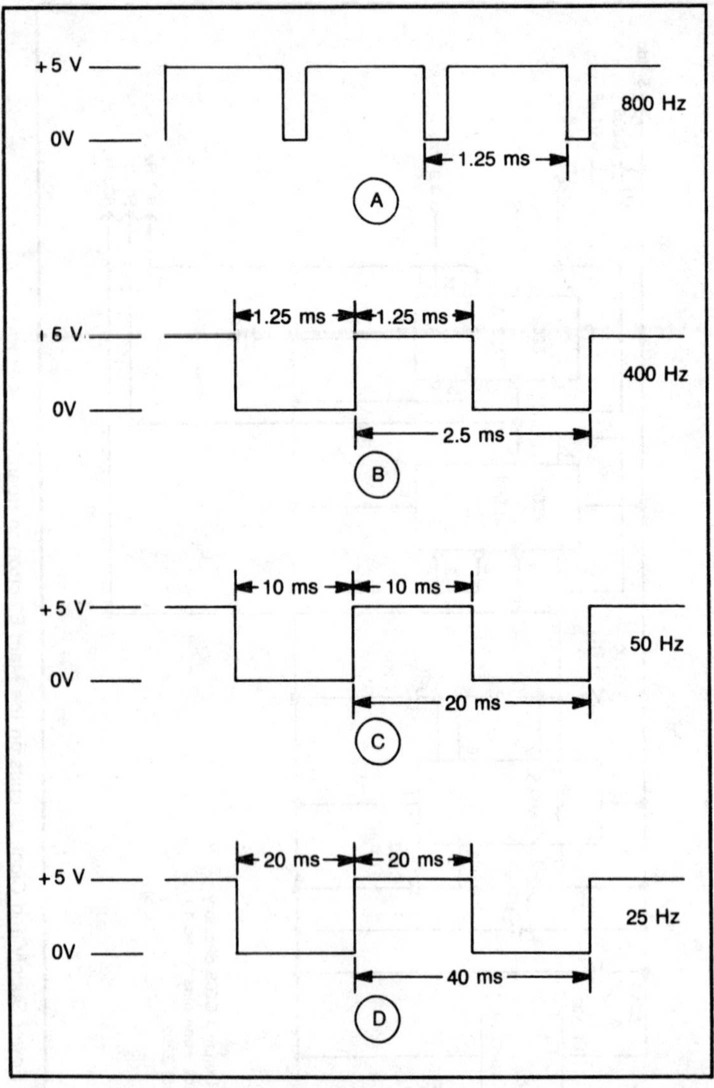

Fig. 4-2A. Waveforms generated by the clocking circuits. (A) Pin 3 of Z200 and pin 14 of Z201. (B) Pins 1 and 12 of Z201. (C) Pin 11 of Z201 and pin 14 of Z202. (D) Pins 1 and 12 of Z202.

tain amount of internal inductance that makes it difficult for them to handle shorter-term voltage spikes. That is why a smaller-valued, nonelectrolytic capacitor, C201, is connected in parallel with it.

The clock-generator circuit consists of an 800 Hz free-running

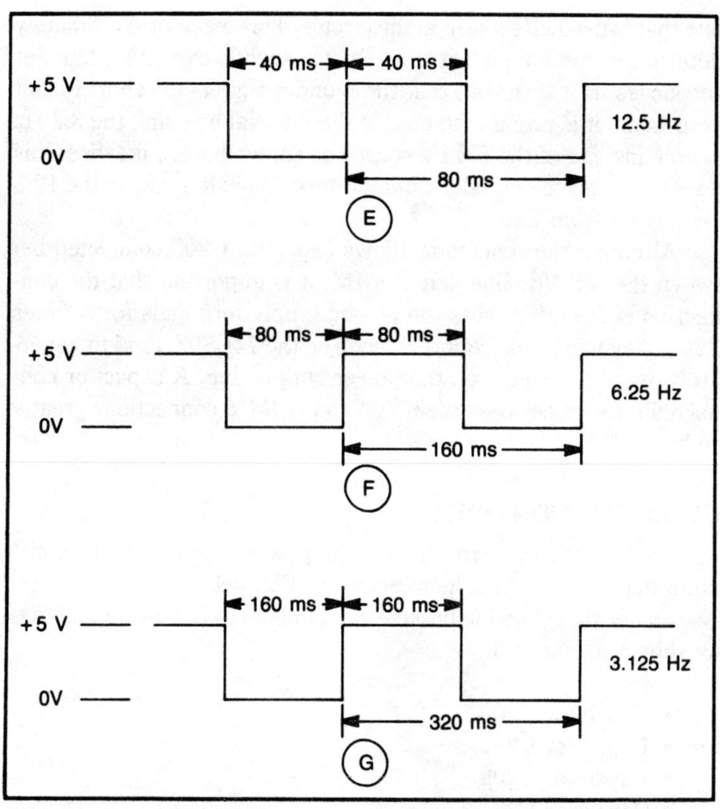

Fig. 4-2B. (E) Pin 9 of Z202. (F) Pin 8 of Z202. (G) Pin 11 of Z202.

multivibrator, Z200, and a pair of cascaded 4-bit binary counters, Z201 and Z202.

The frequency of the multivibrator is determined by the values of R200, R201, and C204. Capacitor C203 simply helps stabilize the operating frequency. Connecting an oscilloscope to the 800 Hz source (between pin 3 of Z200 and COMM), you should see negative-going trigger pulses and a waveform period of about 1.25 milliseconds (see A Fig. 4-2A).

Counter Z201 is triggered directly from the 800 Hz source; and because its output is taken from the high-order bit connection, it generates the symmetrical, 50 Hz waveform illustrated in C Fig. 4-2A.

Counter Z202 is triggered directly from the 50 Hz source. Its four outputs provide symmetrical waveforms of 25 Hz, 12.5 Hz, 6.25 Hz (or about 6.3 Hz) and 3.125 Hz (or about 3.2 Hz).

The schematic indicates that the 6.3 Hz waveform is the only one that is used elsewhere in the circuit. This waveform ultimately determines the amount of time Buster spends executing blunder responses. If it turns out that the blunder time is too short to suit your taste, it is possible to double the interval by using the 3.2 Hz source instead of the 6.3 Hz source as shown here. Likewise, it is possible to cut the blunder timing interval in half by using the 12.5 Hz source from Z202.

Although the schematic shows capacitor C205 connected between the +5 Vdc line and COMM, it is important that the connection is directly across the power supply terminals for counter Z202. Asynchronous counters, such as the 74LS93, tend to generate voltage transients on the power supply line. A capacitor connected directly between their VCC and COMM connections greatly reduces this undesirable effect.

CONSTRUCTION HINTS

Table 4-1 is the parts list for the power supply and clock circuits that are shown schematically in Fig. 4-1.

Solder the following passive components to DIP header CP200 as shown in Fig. 4-3:

- Capacitor C200.
- Capacitor C201.
- Capacitor C203.
- Capacitor C204.
- Resistor R200.
- Resistor R201.

Use Fig. 4-4 as a guide for mounting the following components onto the Main Electronics Board:

- Barrier strip TS200.
- Regulator VR200 and its heat sink.
- Electrolytic capacitor C202.
- 24-pin DIP socket for CK200.
- 8-pin DIP socket for Z200.
- 14-pin DIP socket for Z201.
- 14-pin DIP socket for Z202.

Use the schematic in Fig. 4-1 has a guide for making wire-wrap connections at the IC sockets, solder connections at the voltage regulator, and solderless connections to the terminal strip. Solder

Table 4-1. Parts List for the Power Supply and Clock Portion of the Main Electronics Board.

R200—15 kΩ, 1/4 W resistor
R201—1.8 kΩ, 1/4 W resistor
C200, C201, C204—0.1 μF capacitor
C202—100 μF, 50 WVdc (working voltage dc) electrolytic capacitor
C203, C205—0.01 μF capacitor
VR200—LM309 K positive 5 V regulator
Z200—LM555 timer
Z201, Z202—74LS93 four-bit binary counter
TS200—Two-position, double-row barrier strip (such as Radio Shack #274-656)
CP200—24-pin DIP header
1 TO-220 heat-sink assembly
1 8-pin wire-wrap DIP socket
2 14-pin wire-wrap DIP sockets
1 24-pin wire-wrap DIP socket

capacitor C205 directly between pins 5 and 10 on the socket for Z202.

Insert CP200 and ICs Z200 through Z202 into their designated sockets.

Turn S1, the main power switch off; and use AWG #18 stranded wire to make the following solderless connections between TS200 on the Main Electronics Board and TS3 on the Power Distribution Board:

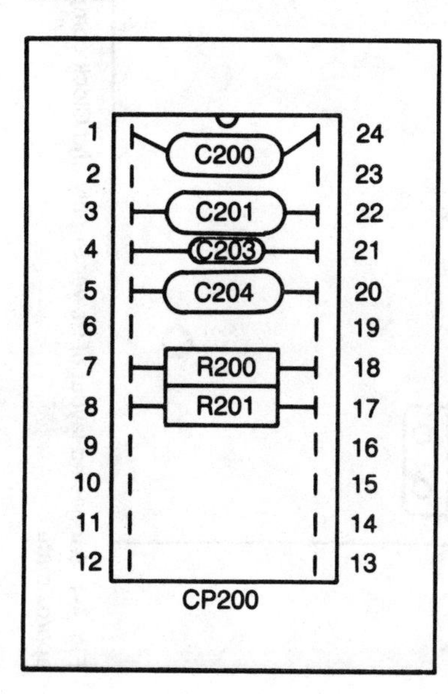

Fig. 4-3. Suggested layout of Power Supply and Clock components for the 24-pin DIP header, CJ200.

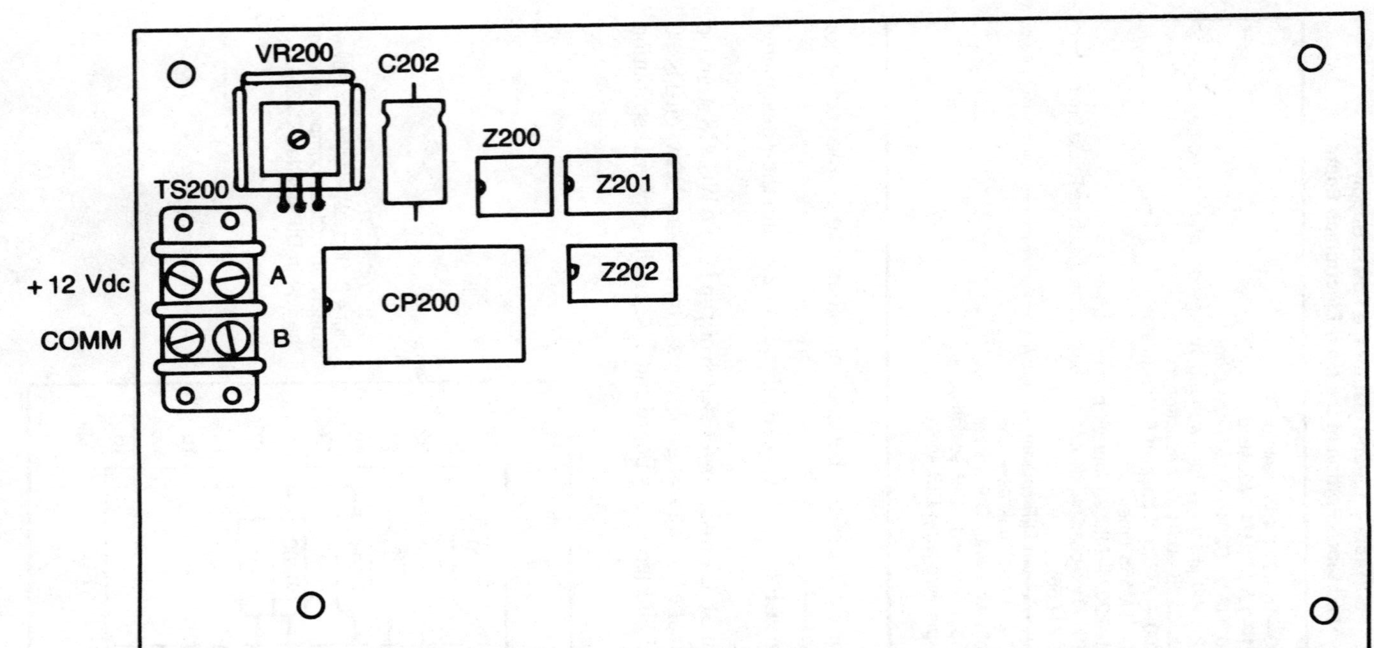

Fig. 4-4. Suggested layout of Power Supply and Clock components on the Main Electronics Board. The dimensions and proportions are approximate.

- +12 Vdc connection between TS3-C and TS200-A.
- COMM connection between TS3-F and TS200-B.

TESTING HINTS

Turn the main power switch S1, on, and conduct the following tests to confirm the operation of the voltage regulator and proper distribution of power to the IC devices:

- Measure +12 Vdc between terminals A and B on TS200.
- Measure +12 Vdc on the IN pin of VR200.
- Measure +5 Vdc (plus or minus 0.5 V) at the OUT pin of VR200.
- Measure about +5 Vdc at pin 8 of Z200.
- Measure about +5 Vdc at pin 5 of Z201.
- Measure about +5 Vdc at pin 5 of Z202.

Use the following procedures to check the operation of the clock circuitry:

- Note the waveform in Fig. 4-2A at pin 3 of Z200, or measure approximately +4 Vdc at that point.
- Note the waveform in Fig. 4-2A at pin 14 of Z201, or measure approximately +4 Vdc at that point.
- Note the waveform in Fig. 4-2B at pins 1 and 12 of Z201, or measure approximately 2.5 Vdc at those points.
- Note the waveform in Fig. 4-2C at pin 11 of Z201, or measure approximately 2.5 Vdc at that point.
- Note the waveform in Fig. 4-2C at pin 14 of Z202, or measure approximately 2.5 Vdc at that point.
- Note the waveform in Fig. 4-2D at pins 1 and 12 of Z202, or measure approximately 2.5 Vdc at those points.
- Note the waveform in Fig. 4-2E at pin 9 of Z202, or measure approximately 2.5 Vdc at that point.
- Note the waveform in Fig. 4-2F at pin 8 of Z202, or measure approximately 2.5 Vdc at that point.
- Note the waveform in Fig. 4-2G at pin 11 of Z202, or measure approximately 2.5 Vdc at that point.

Chapter 5

The Blunder Input Interface Circuit

The primary purpose of the Blunder Input circuitry is to provide a stable source of binary information regarding the status of Buster's blunder switches. The circuit continually monitors the status of these switches and updates the 4-bit blunder bus at 20-millisecond intervals.

CIRCUIT DESCRIPTION

Figure 5-1 is the schematic diagram of the Blunder Input Interface portion of the Main Electronics Board. The circuit is connected directly to the blunder switches through connections BS0 through BS3 at terminal strip TS201.

Resistors R202 through 205 are connected between these inputs and +5 Vdc in order to pull up the signal levels. Using this scheme, open switches generate logic-1 levels, while closed switches generate logic-0 levels. Table 5-1 summarizes the logic levels at inputs BS0, BS1, BS2 and BS3 as a function of all 16 possible combinations of the blunder-switch conditions.

Whenever there is no blunder contact, all four input levels are in logic 1. A front-only contact sets BS0 to logic 0 but leaves the others at logic 1. Each kind of blunder condition generates a distinctive 4-bit pattern.

Under ideal operating conditions, the circuit as described thus far would be sufficient for feeding blunder contact codes to other parts of the system that use the information. Buster does not operate under ideal conditions, however. In fact, the blunder switches tend to be very noisy. They generate a rapid series of voltage spikes for a period of 5 to 10 milliseconds whenever a switch is pushed closed. This phenomenon, called *switch contact bounce*, must be eliminated or greatly reduced. This is the primary purpose of the remaining portion of the circuit.

One of the key elements in the switch-debouncing scheme is the 4-bit data latch, Z208. Notice that inputs BS0 through BS3 are connected directly to the data inputs of this device. The latch works something like a simple, 4-bit memory circuit. The circuit can read the status of the blunder switches only as long as the gating signal at pins 4 and 13 is at a logic-1 level. Once that gating signal goes to logic 0, the circuit remembers the 4-bit pattern and ignores any further changes in the input logic levels.

The general idea is to gate the latch, or update its outputs, at times when there is not likely to be a great deal of contact bouncing taking place. When the gate is latched, blunder bus (BB0 through BB3) and the inputs to NAND gate Z207-B rarely allow contact-bounce noise to subsequent sections of the system.

The gating waveform applied to pins 4 and 13 of the latch is a 1.25 millisecond, positive-going pulse. The pulse normally occurs at a regular, 50 Hz rate. This means the circuit updates the latch data 50 times a second. But more important, there is always a 20-millisecond delay whenever the blunder status changes from NO CONTACT to any contact conditions. Likewise, there is a 20-millisecond delay whenever the status changes from a contact to a NO CONTACT condition. This delay—an interruption in the normal 50 Hz updating rate—is wholly responsible for eliminating contact-bounce noise at the most critical times.

A 4-input NAND gate, Z203-A, senses any contact at the blunder switch inputs. Because the Blunder Input uses an active-low format (1 = no contact, 0 = contact) the device actually behaves as an OR gate. The output of Z203-A is thus fixed at logic 0 as long as none of the blunder switches is closed. It rises to logic 1 whenever any combination of blunder switches is closed.

The output of Z203-A is connected directly to the reset terminals of a 4-bit binary counter, Z206. And it is also connected through a NAND gate, Z204-A, to the reset terminals of a second binary counter, Z205.

Both counters are clocked with the 800 Hz waveform shown in Fig. 5-2A. But the connections from Z203-A and Z204-A are arranged so that one counter is always running while the other is not.

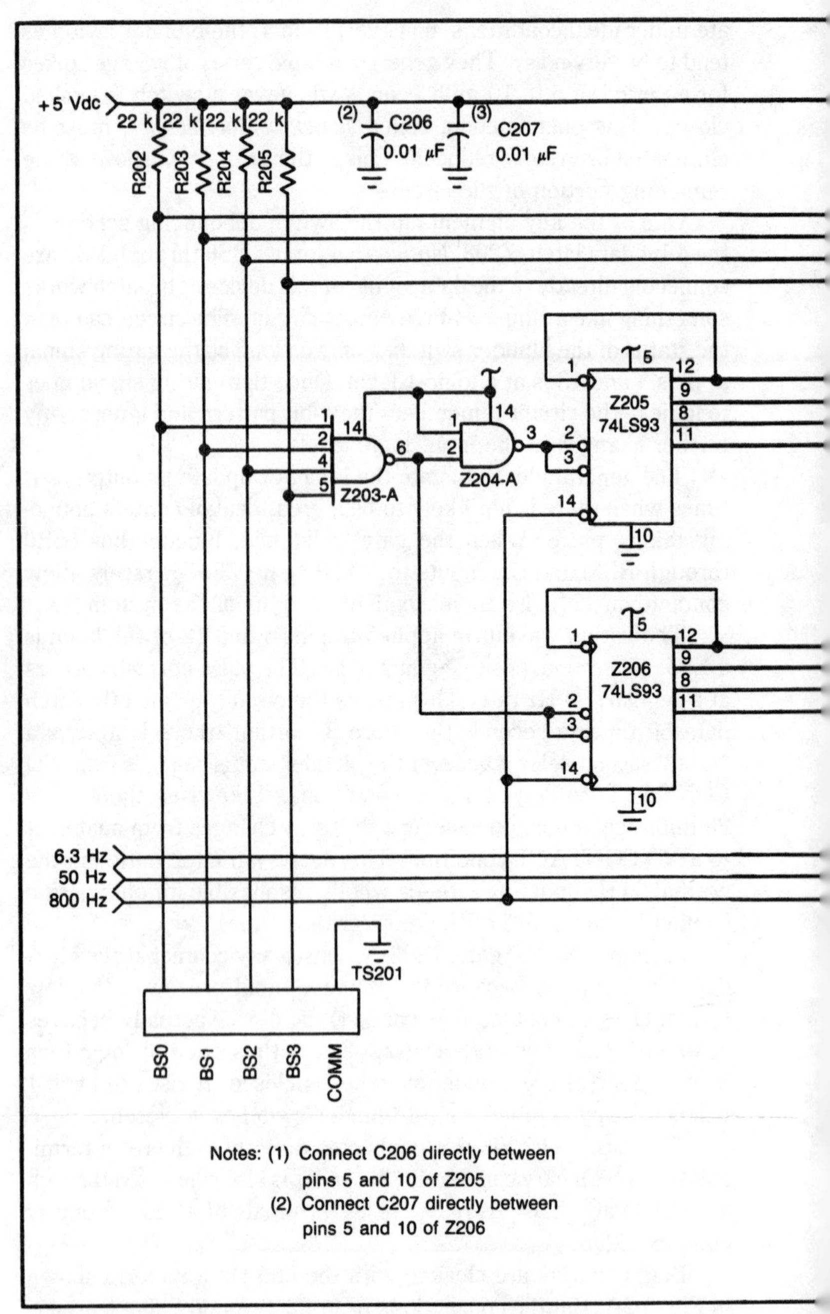

Fig. 5-1. Schematic of the Blunder Input interface portion of the Main Electronics Board.

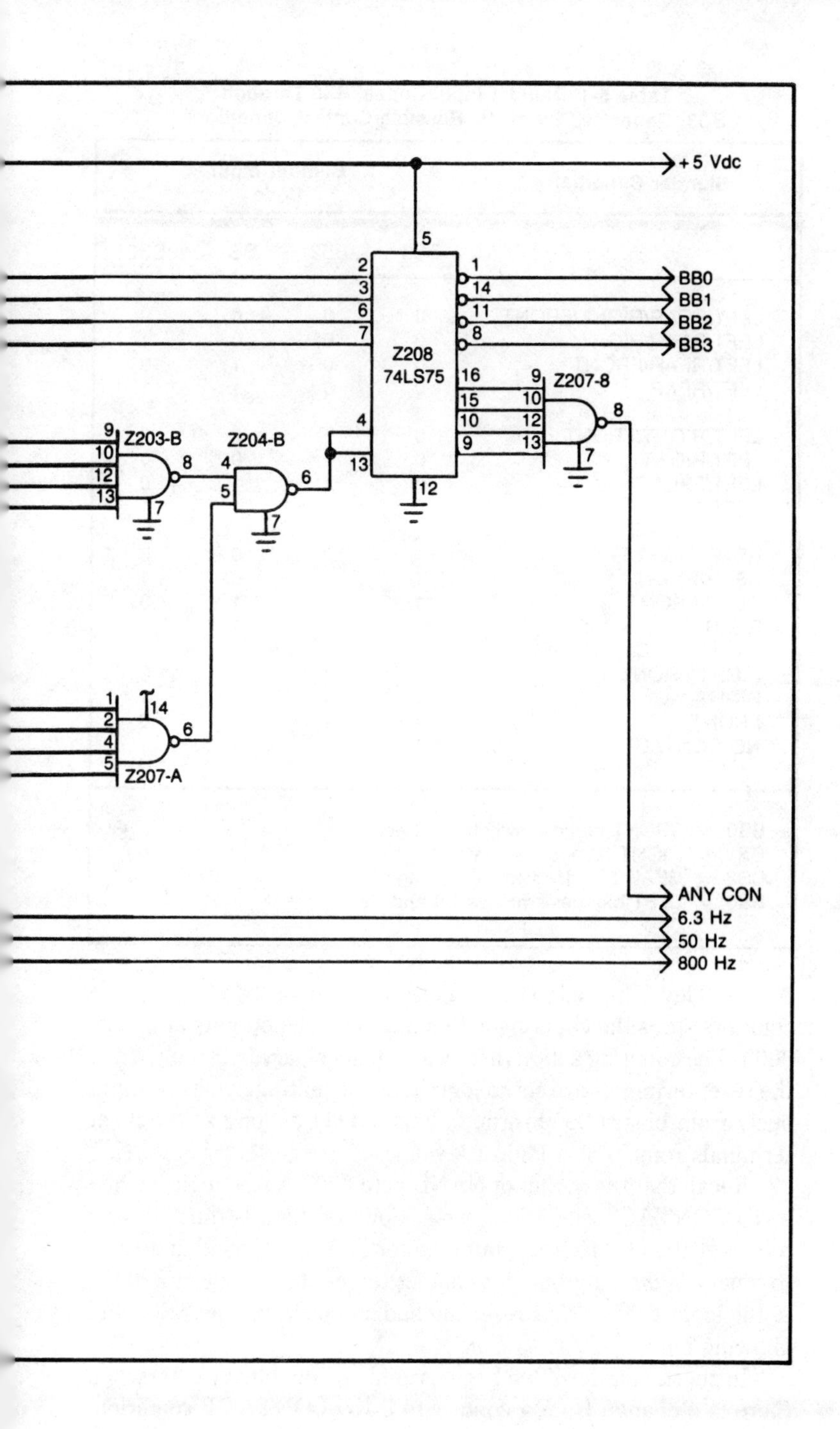
+5 Vdc
Z208
74LS75
BB0
BB1
BB2
BB3
Z207-8
Z203-B
Z204-B
Z207-A
ANY CON
6.3 Hz
50 Hz
800 Hz

Table 5-1. Blunder Input Codes, BS0 Through BS3, Generated By All 16 Possible Contact Conditions.

Blunder Condition	Blunder Input Code			
	BS3	BS2	BS1	BS0
LEFT/REAR/RIGHT/FRONT	0	0	0	0
LEFT/REAR/RIGHT	0	0	0	1
LEFT/REAR/FRONT	0	0	1	0
LEFT/REAR	0	0	1	1
LEFT/RIGHT/FRONT	0	1	0	0
LEFT/RIGHT	0	1	0	0
LEFT/FRONT	0	1	1	0
LEFT	0	1	1	1
REAR/RIGHT/FRONT	1	0	0	0
REAR/RIGHT	1	0	0	1
REAR/FRONT	1	0	1	0
REAR	1	0	1	1
RIGHT/FRONT	1	1	0	0
RIGHT	1	1	0	1
FRONT	1	1	1	0
NO CONTACT	1	1	1	1

BS0 = FRONT blunder switches S2 and S3
BS1 = RIGHT blunder switches S4 and S5
BS2 = REAR blunder switches S6 and S7
BS3 = LEFT blunder switches S8 and S9

A logic-0 level applied to the reset terminals (pins 2 and 3) on these counters stops the counting action and fixes the outputs at binary 0000. The counting action then begins from binary 0000 only when the reset terminals are set to logic 1. The counting cycles continuously from binary 0000 through binary 1111 as long as the reset terminals remain at a logic-1 level.

Recall that the output of NAND gate Z203-A is at logic 0 while a NO CONTACT condition exists. Counter Z206 is thus allowed to run at the 800 Hz rate, but counter Z205 is stopped and reset to binary 0000. Any blunder contact forces the output of Z203-A to the logic-1 level, thus resetting and stopping counter Z206 and allowing counter Z205 to run.

In short, counter Z206 begins running from binary 0000 when there is a change from a contact to a NO CONTACT condition.

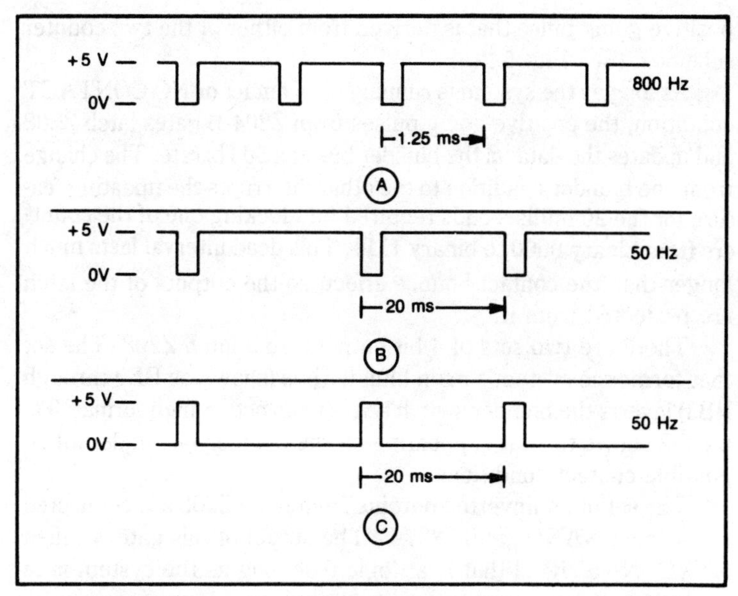

Fig. 5-2. Waveforms used by the Blunder Input interface circuits. (A) Pin 14 of Z205 and Z206. (B) Pin 6 of Z207-A when no blunder switch is closed. (C) Pin 6 of Z204-B, and pins 4 and 13 of Z208.

Counter Z205 begins running from binary 0000 when there is a change from the NO CONTACT to any contact condition.

The four outputs from both counters are connected to 4-input NAND gates. Z203-B is connected to the outputs of counter Z205, and Z207-A is connected to the inputs of counter Z206. The output of these gates is normally at logic 1, and drop to logic 0 only while the corresponding counter is at binary 1111.

Suppose that a NO CONTACT condition exists. This means that the output of Z203-A is fixed at the logic-0 level and Z206 is continuously cycling through its 16 states at an 800 Hz rate. The output of Z207-A is normally at logic 1, falling to logic 0 only during the time the counter is at binary 1111—for 1.25 milliseconds at 20-millisecond intervals. Counter Z205 and gate Z203-B generate the same sort of waveform, but only while any sort of contact condition exists. (See Fig. 5-2B.)

The outputs of gates Z207-A and Z203-B use an active-low format. That is, their outputs are normally at logic 1, and fall to logic 0 whenever they are activated by the binary 1111 output from their counter. The active-low format makes it possible to use NAND gate Z204-B as an OR circuit, so the output from Z204-B is a 50 Hz, positive-going pulse that is derived from either of the two counter schemes. (See Fig. 5-2C.)

As long as the system is running in a contact or NO CONTACT condition, the positive-going pulses from Z204-B gates latch Z208 and updates the data on the blunder bus at a 50 Hz rate. The change from one blunder condition to the other interrupts the updating feature for the 20 milliseconds required for clocking one of the counters from binary 0000 to binary 1111. This dead interval lasts much longer than the contact-bounce effect, so the outputs of the latch are protected from it.

There are two sets of 4-bit outputs from latch Z208. The set that forms the system's main blunder bus (shown as BB0 through BB3) inverts the blunder switch levels to an active-high format. Table 5-2 shows the data appearing on the blunder bus under all 16 possible contact conditions.

The set of noninverted outputs from latch Z208 are connected to a 4-input NAND gate, Z207-B. The output of this gate is called ANYCON—a signal that is at logic 0 as long as the system is in

Table 5-2. Blunder-Buss Codes, BB0 Through BB3, Generated By All 16 Possible Contact Conditions.

Blunder Condition	Blunder Bus Code			
	BB3	BB2	BB1	BB0
LEFT/REAR/RIGHT/FRONT	1	1	1	1
LEFT/REAR/RIGHT	1	1	1	0
LEFT/REAR/FRONT	1	1	0	1
LEFT/REAR	1	1	0	0
LEFT/RIGHT/FRONT	1	0	1	1
LEFT/RIGHT	1	0	1	0
LEFT/FRONT	1	0	0	1
LEFT	1	0	0	0
REAR/RIGHT/FRONT	0	1	1	1
REAR/RIGHT	0	1	1	0
REAR/FRONT	0	1	0	1
REAR	0	1	0	0
RIGHT/FRONT	0	0	1	1
RIGHT	0	0	1	0
FRONT	0	0	0	1
NO CONTACT	0	0	0	0

Table 5-3. Parts List for the Blunder Input Interface Portion of the Main Electronics Board.

R202, R203, R204, R205—22 kΩ, 1/4 W resistor
C206, C207—0.01 μF capacitor
Z203, Z207—74LS40 dual 4-input NAND gate
Z204—74LS00 quad 2-input NAND gate
Z205, Z206—74LS93 4-bit latch
TS201—6-position, dual-row barrier strip (such as Radio Shack #274-659)
5 ea. 14-pin wire-wrap DIP socket
1 ea. 16-pin wire-wrap DIP socket

the NO CONTACT mode, and rises to logic 1 while any valid contact condition exists.

CONSTRUCTION HINTS

Table 5-3 is the parts list for the blunder input interface circuit that is shown schematically in Fig. 5-1.

Solder the following passive components to DIP header CP200 as shown in Fig. 5-3:

- Resistor R202
- Resistor R203
- Resistor R204
- Resistor R205

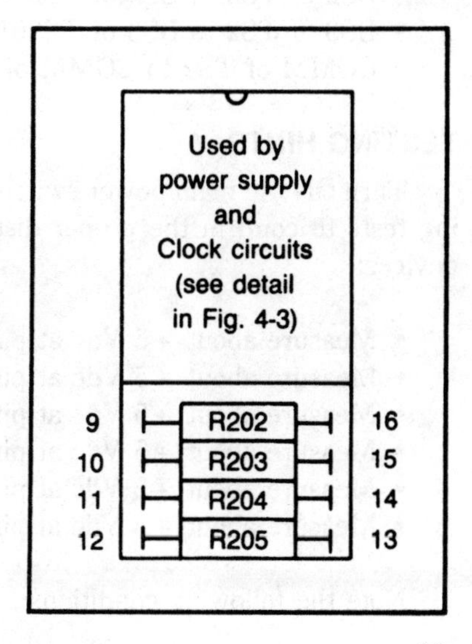

Fig. 5-3. Suggested layout of Blunder Input interface components on the 24-pin DIP header, CJ200.

Use Fig. 5-4 as a guide for mounting the following components onto the Main Electronics Board:

- Barrier strip TS201
- 14-pin DIP socket for Z203
- 14-pin DIP socket for Z204
- 14-pin DIP socket for Z205
- 14-pin DIP socket for Z206
- 14-pin DIP socket for Z207
- 16-pin DIP socket for Z208

Use the schematic in Fig. 5-1 as a guide for making all wire-wrap, solder and solderless connections. Also:

- Solder C206 directly between pins 5 and 10 of Z205.
- Solder C207 directly between pins 5 and 10 of Z206.

Make sure the main power switch is turned OFF, then insert ICs Z203 through Z208 into their designated sockets.

Make the following solderless connections between the blunder switch terminal strip, TS2, and the Blunder Input terminal block, TS201, on the Main Electronics Board:

- BS0 of TS2 to BS0 of TS201
- BS1 of TS2 to BS1 of TS201
- BS2 of TS2 to BS2 of TS201
- BS3 of TS2 to BS3 of TS201
- COMM of TS2 to COMM of TS201

TESTING HINTS

Turn ON the main power switch, S1, and conduct the following tests to confirm the proper distribution of power to the IC devices:

- Measure about +5 Vdc at pin 14 of Z203.
- Measure about +5 Vdc at pins 1 and 14 of Z204.
- Measure about +5 Vdc at pin 5 of Z205.
- Measure about +5 Vdc at pin 5 of Z206.
- Measure about +5 Vdc at pin 14 of Z207.
- Measure about +5 Vdc at pin 5 of Z208.

Note the following conditions:

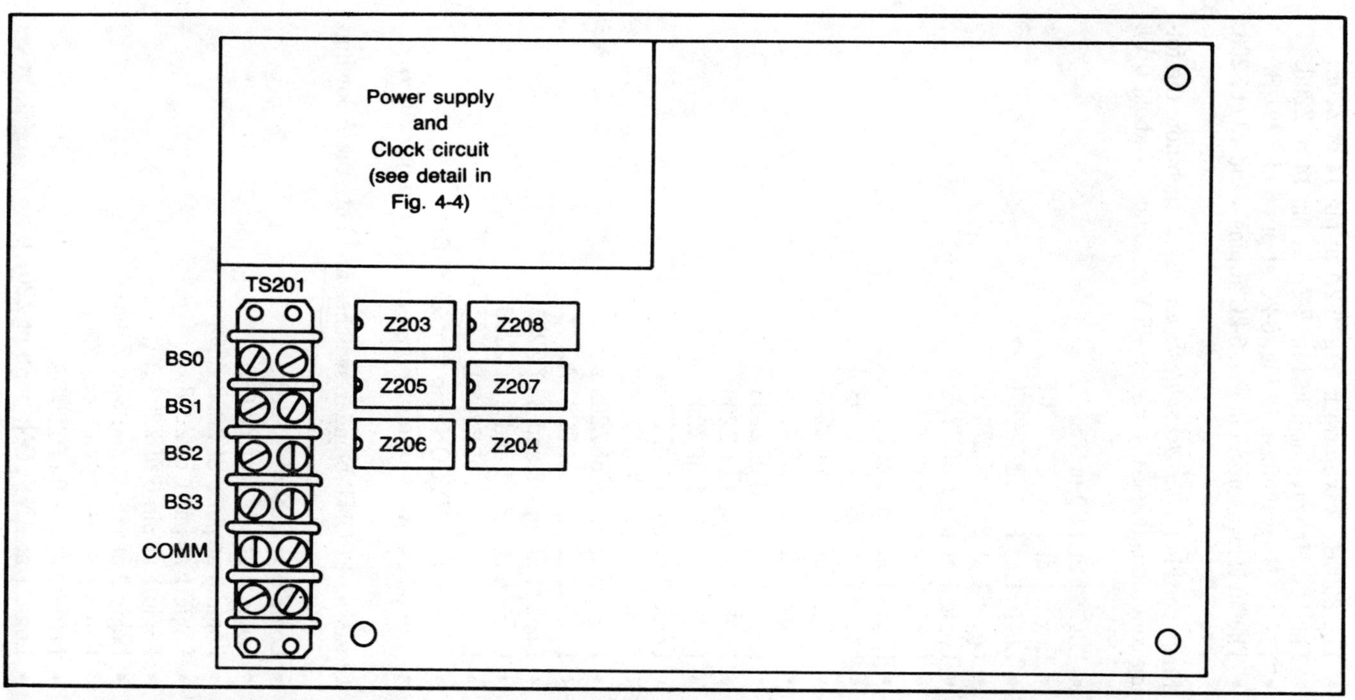

Fig. 5-4. Suggested layout of the Blunder Input interface components on the Main Electronics Board.

- The 800 Hz waveform in Fig. 5-2A at pin 14 of Z205.
- The 800 Hz waveform in Fig. 5-2A at pin 14 of Z206.
- The 50 Hz waveform in Fig. 5-2C at pin 6 of Z204-B.
- The 50 Hz waveform in Fig. 5-2C at pins 4 and 13 of Z208.

Make certain *no blunder switches are closed*, then note the following conditions (logic 1 = about +5 Vdc, logic 0 = about 0 V):

- Logic 1 at BS0 of TS201
- Logic 1 at BS1 of TS201
- Logic 1 at BS2 of TS201
- Logic 1 at BS3 of TS201
- Logic 1 at pin 1 of Z203-A
- Logic 1 at pin 2 of Z203-A
- Logic 1 at pin 4 of Z203-A
- Logic 1 at pin 5 of Z203-A
- Logic 1 at pin 2 of Z208
- Logic 1 at pin 3 of Z208
- Logic 1 at pin 6 of Z208
- Logic 1 at pin 7 of Z208
- Logic 0 at pin 1 of Z208
- Logic 0 at pin 14 of Z208
- Logic 0 at pin 11 of Z208
- Logic 0 at pin 8 of Z208
- Logic 0 at pin 6 of Z203-A
- Logic 1 at pins 2 and 3 of Z205
- Logic 0 at pins 2 and 3 of Z206
- Logic 1 at pin 8 of Z203-B
- Logic 0 at ANYCON, pin 8 of Z207-B
- The 50 Hz waveform of Fig. 5-2B at pin 6 of Z207-A

Close the FRONT blunder switches and note the following:

- Logic 0 at BS0 of TS201
- Logic 0 at pin 1 of Z203-A
- Logic 0 at pin 2 of Z208
- Logic 1 at pin 1 of Z208
- Logic 1 at pin 6 of Z203-A
- Logic 0 at pins 2 and 3 of Z205
- Logic 1 at pins 2 and 3 of Z206
- Logic 1 at pin 6 of Z207-A
- Logic 1 at ANYCON, pin 8 of Z207-B

- The 50 Hz waveform in Fig. 5-2B at pin 8 of Z203-B

Close the RIGHT blunder switches and note the following:

- Logic 0 at BS1 of TS201
- Logic 0 at pin 2 of Z203-A
- Logic 0 at pin 3 of Z208
- Logic 1 at pin 14 of Z208

Close the REAR blunder switches and note the following:

- Logic 0 at BS2 of TS201
- Logic 0 at pin 4 of Z203-A
- Logic 0 at pin 3 of Z208
- Logic 1 at pin 11 of Z208

Close the LEFT blunder switches and note the following:

- Logic 0 at BS3 of TS201
- Logic 0 at pin 5 of Z203-A
- Logic 0 at pin 7 of Z208
- Logic 1 at pin 8 of Z208

Chapter 6

The Blunder Response Encoder

Three out of four of Buster's operating modes requires him to make automatic blunder responses to contact conditions. The purpose of the Blunder Response Encoder is to transform the current data on the blunder bus into a 4-bit response for the D/S motor system. In a manner of speaking, the circuit represents Buster's programmed responses to all possible blunder conditions.

CIRCUIT DESCRIPTION

The Blunder Condition column in Table 6-1 specifies all possible blunder conditions, and the Blunder Bus Code column shows the 4-bit blunder bus codes associated with each of them. The blunder bus data is generated by the interface circuitry described in Chapter 5.

The two remaining columns indicate Buster's programmed response to each of the blunder conditions. The Blunder Response column describes the platform motions that are to take place, and the Motor Bus Code column shows the 4-bit motor-control codes associated with each response.

So when Buster is running freely across the floor, he is in a NO CONTACT blunder condition. The table suggests that the appropriate motor response is to continue running FWD (forward). But when he brushes against an object on the right side, the blunder condition is RIGHT, and the table suggests that the proper mo-

Table 6-1. Blunder Conditions and Motor Response Codes Generated by the Blunder Response Encoder Circuitry.

Blunder Condition	Blunder Bus Code				Robot Response	Motor Bus Code			
	BB3	BB2	BB1	BB0		P1D	P1C	P1B	P1A
NO CONTACT	0	0	0	0	FWD	0	0	0	0
FRONT	0	0	0	1	REV RIGHT	0	1	0	1
RIGHT	0	0	1	0	FWD LEFT	0	0	1	0
RIGHT/FRONT	0	0	1	1	REV LEFT	0	1	1	0
REAR	0	1	0	0	FWD	0	0	0	0
REAR/FRONT	0	1	0	1	CW SPIN	0	0	1	1
REAR/RIGHT	0	1	1	0	FWD LEFT	0	0	1	0
REAR/RIGHT/FRONT	0	1	1	1	CCW SPIN	0	1	1	1
LEFT	1	0	0	0	FWD RIGHT	0	0	0	1
LEFT/FRONT	1	0	0	1	REV RIGHT	0	1	0	1
LEFT/RIGHT	1	0	1	0	FWD	0	0	0	0
LEFT/RIGHT/FRONT	1	0	1	1	REV	0	1	0	0
LEFT/REAR	1	1	0	0	FWD RIGHT	0	0	0	1
LEFT/REAR/FRONT	1	1	0	1	CW SPIN	0	0	1	1
LEFT/REAR/RIGHT	1	1	1	0	FWD	0	0	0	0
LEFT/REAR/RIGHT/FRONT	1	1	1	1	STOP	1	1	1	1

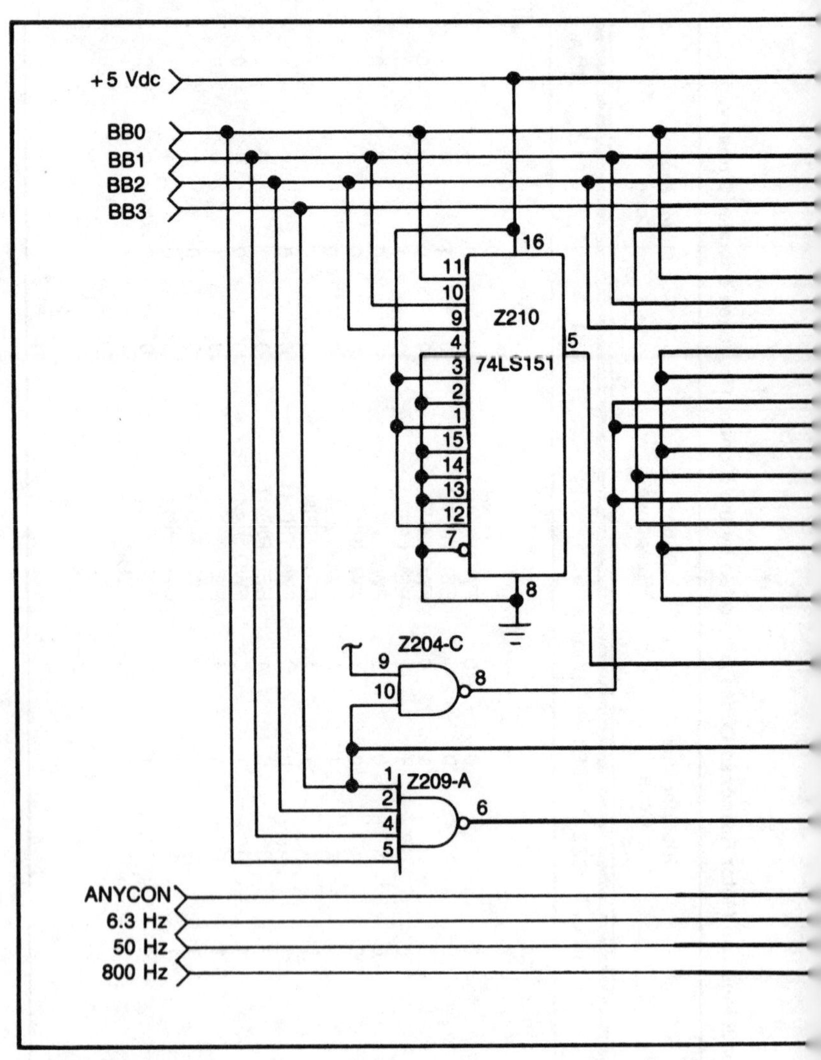

Fig. 6-1. Schematic of the Blunder-Response Encoder circuitry on the Main Electronics Board.

tor response should be FWD LEFT (forward with a left turn). Study the table thoroughly, noting the responses assigned to each of the blunder conditions.

The purpose of the blunder response encoder, shown schematically in Fig. 6-1, is to transform the 4-bit blunder bus codes into corresponding motor bus codes. Z212 translates the blunder codes into bit P1A, Z211 does the job for bit P1B, and Z210 does the con-

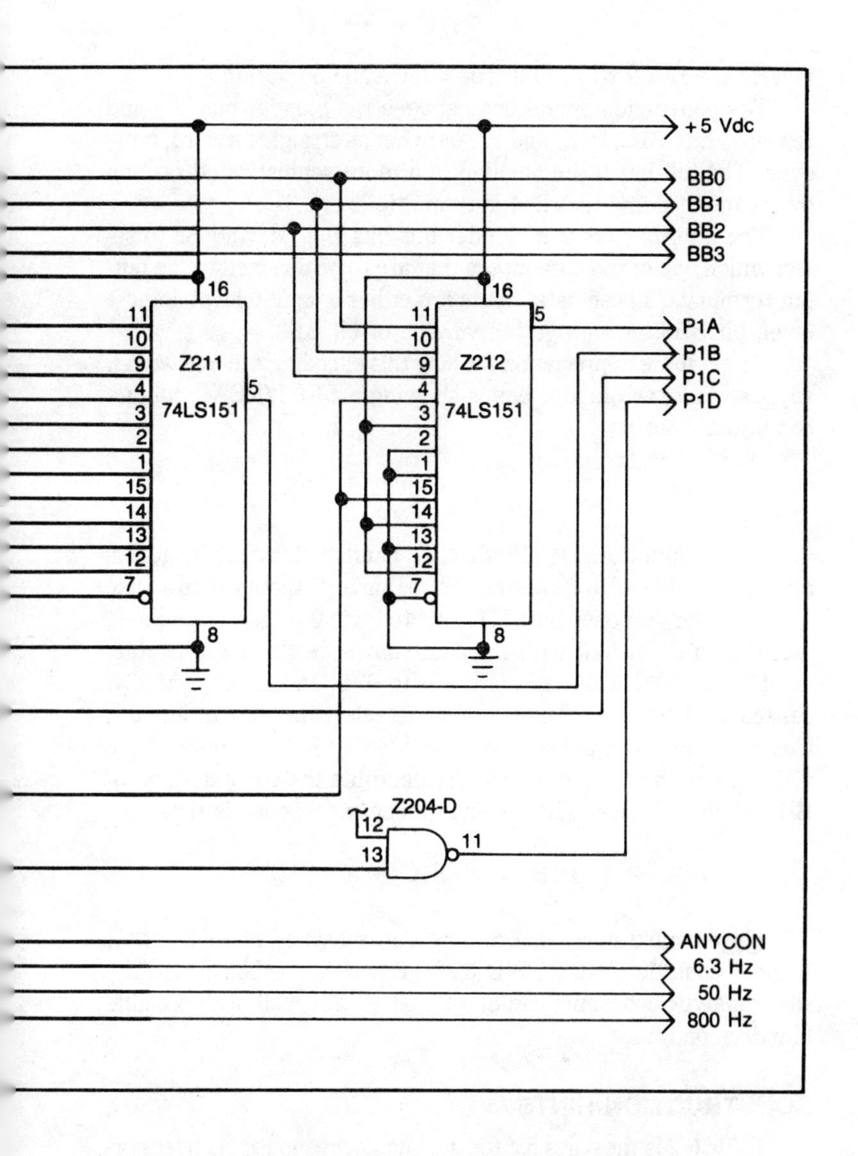

version for bit P1C. NAND gates Z209-A and Z204-D work together to translate the blunder bus bits into response bit P1D.

The table shows that response bit P1D is equal to 0 under all blunder conditions except on—where the blunder code is binary 1111. The code encoding task for response bit P1D is thus a simpler matter of ANDing the four bits on the blunder bus. Z209-A executes a NAND operation on the blunder bus bits, and Z204-D inverts the result to yield a true 4-bit AND operation.

The conversion operations between the blunder bus bits and response bits P1C, P1B, and P1A are not as straightforward, however. The job has to be handled in a more sophisticated fashion by means of 8-line to 1-line data multiplexers.

The multiplexers use blunder bus bits B0, B1, and B2 to select which one of the data inputs that are to be directed to the output terminals. These data inputs are either a logic-0 level, logic-1 level, bit BB3, or an inverted version of bit BB3.

To get more appreciation for how this encoding scheme works, suppose that the blunder bus is showing a LEFT/REAR contact condition. That is,

BB0 = 0 BB1 = 0 BB2 = 1 BB3 = 1

The 4-input NAND gate, Z209-A, handles this code by generating a logic-1 level at its output. Z204-D inverts that level to a logic 0, so blunder response bit P1D is set to logic 0 in this example—a fact that is consistent with the data shown in the encoder table.

Having BB0, BB1, and BB2 set to 0, 0, and 1, respectively, causes all three multiplexer devices to select the data at input 4. Counting the first input as 0, you can see that Z210 outputs a logic 0, Z211 outputs a logic 0, and Z212 outputs the current value of BB3—1 in this case. The encoded blunder response is thus:

P1A = 1 P1B = 0 P1C = 0 P1D = 0

The example shows that the encoder translates a LEFT/REAR blunder condition into a FWD RIGHT blunder response, and the same analytic procedure can be applied equally well to all possible blunder conditions.

CONSTRUCTION HINTS

Table 6-2 is the parts list for the blunder response encoder portion of the Main Electronics Board that is illustrated in Fig. 6-1.

Use Fig. 6-2 as a guide for mounting the following components onto the Main Electronics Board:

- 14-pin DIP socket for Z209
- 16-pin DIP socket for Z210
- 6-pin DIP socket for Z211
- 16-pin DIP socket for Z212

Table 6-2. Parts List for the Blunder Encoder Portion of the Main Electronics Board.

Z209—74LS40 4-input NAND gate
Z210, Z211, Z212—74LS151 8-line to 1-line data multiplexer
1 ea. 14-pin wire-wrap DIP socket
3 ea. 16-pin wire-wrap DIP socket

Use the schematic in Fig. 6-1 as a guide for making all wire-wrap connections.

Make certain that the main power switch is off, then insert ICs Z209 through Z212 into their designated sockets.

TESTING HINTS

Turn ON the main power switch, and conduct the following tests to confirm the proper distribution of power to the IC devices:

- Measure about +5 Vdc at pin 14 of Z209
- Measure about +5 Vdc at pin 16 of Z210
- Measure about +5 Vdc at pin 16 of Z211
- Measure about +5 Vdc at pin 16 of Z212

Make certain that all blunder switches are open (NO CONTACT blunder condition), and note the FWD response code as follows:

- Logic 0 at pin 11 of Z204-D
- Logic 0 at pin 5 of Z210
- Logic 0 at pin 5 of Z211
- Logic 0 at pin 5 of Z212

Close only the RIGHT blunder switches (RIGHT blunder condition), and note the FWD LEFT response code as follows:

- Logic 0 at pin 11 of Z204-D
- Logic 0 at pin 5 of Z210
- Logic 1 at pin 5 of Z211
- Logic 0 at pin 5 of Z212

Close the RIGHT and FRONT blunder switches

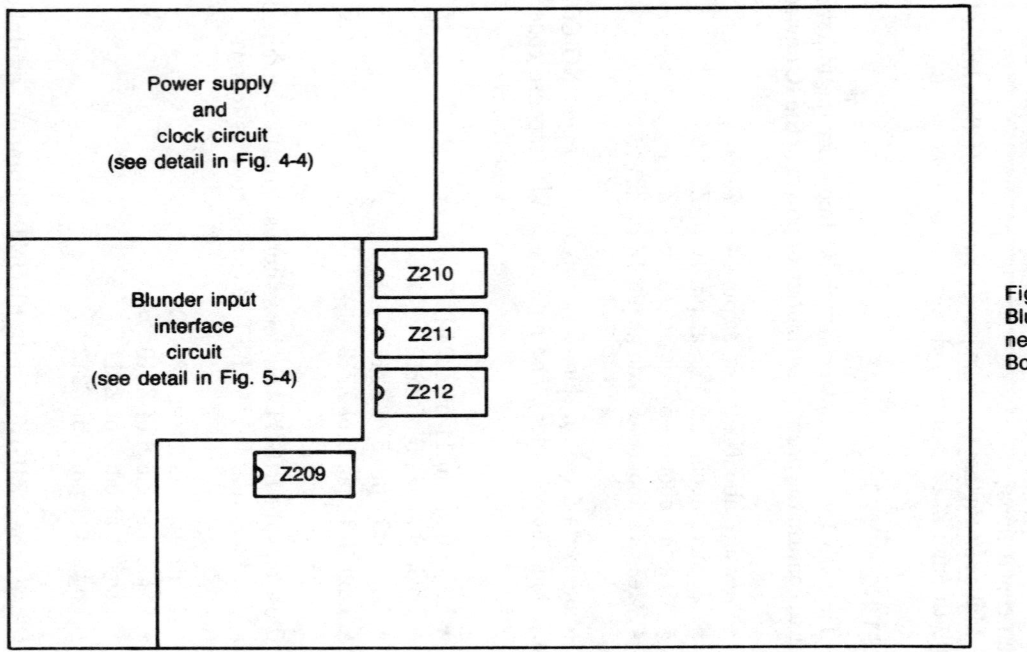

Fig. 6-2. Suggested layout of Blunder-Response Encoder components on the Main Electronics Board.

(RIGHT/FRONT blunder condition), and note the REV LEFT response code as follows:

- Logic 0 at pin 11 of Z204-D
- Logic 1 at pin 5 of Z210
- Logic 1 at pin 5 of Z211
- Logic 0 at pin 5 of Z212

Close only the REAR blunder switches (REAR blunder condition), and note the FWD response code as follows:

- Logic 0 at pin 11 of Z204-D
- Logic 0 at pin 5 of Z210
- Logic 0 at pin 5 of Z211
- Logic 0 at pin 5 of Z212

Close the REAR and FRONT blunder switches (REAR/FRONT blunder condition), and note the CW SPIN response code as follows:

- Logic 0 at pin 11 of Z204-D
- Logic 0 at pin 5 of Z210
- Logic 1 at pin 5 of Z211
- Logic 1 at pin 5 of Z210

Close the REAR and RIGHT blunder switches (REAR/RIGHT blunder condition), and note the FWD LEFT response code as follows:

- Logic 0 at pin 11 of Z204-D
- Logic 0 at pin 5 of Z210
- Logic 1 at pin 5 of Z211
- Logic 0 at pin 5 of Z212

Close the REAR, RIGHT, and FRONT blunder switches (REAR/RIGHT/FRONT blunder condition), and note the CCW SPIN response code as follows:

- Logic 0 at pin 11 of Z204-D
- Logic 1 at pin 5 of Z212
- Logic 1 at pin 5 of Z211
- Logic 1 at pin 5 of Z212

Close only the LEFT blunder switches (LEFT blunder condition), and note the FWD RIGHT response code as follows:

- Logic 0 at pin 11 of Z204-D
- Logic 0 at pin 5 of Z210
- Logic 0 at pin 5 of Z211
- Logic 1 at pin 5 of Z212

Close the LEFT and FRONT blunder switches (LEFT/FRONT blunder condition), and note the REV RIGHT response code as follows:

- Logic 0 at pin 11 of Z204-D
- Logic 1 at pin 5 of Z210
- Logic 0 at pin 5 of Z211
- Logic 1 at pin 5 of Z212

Close the LEFT and RIGHT blunder switches (LEFT/RIGHT blunder condition), and note the FWD response code as follows:

- Logic 0 at pin 11 of Z204-D
- Logic 0 at pin 5 of Z210
- Logic 0 at pin 5 of Z211
- Logic 0 at pin 5 of Z212

Close the LEFT, RIGHT, and FRONT blunder switches (LEFT/RIGHT/FRONT blunder condition), and note the REV response code as follows:

- Logic 0 at pin 11 of Z204-D
- Logic 1 at pin 5 of Z210
- Logic 0 at pin 5 of Z211
- Logic 0 at pin 5 of Z212

Close the LEFT and REAR blunder switches (LEFT/REAR blunder condition), and note the FWD RIGHT response code as follows:

- Logic 0 at pin 11 of Z204-D
- Logic 0 at pin 5 of Z210
- Logic 0 at pin 5 of Z211
- Logic 1 at pin 5 of Z212

Close the LEFT, REAR, and FRONT blunder switches (LEFT/REAR/FRONT blunder condition), and note the CW SPIN response code as follows:

- Logic 0 at pin 11 of Z204-D
- Logic 0 at pin 5 of Z210
- Logic 1 at pin 5 of Z211
- Logic 1 at pin 5 of Z212

Close the LEFT, REAR, and RIGHT blunder switches (LEFT/REAR/RIGHT blunder condition), and note the FWD response code as follows:

- Logic 0 at pin 11 of Z204-D
- Logic 0 at pin 5 of Z210
- Logic 0 at pin 5 of Z211
- Logic 0 at pin 5 of Z212

Close all blunder switches (LEFT/REAR/RIGHT/FRONT blunder condition), and note the STOP response code as follows:

- Logic 1 at pin 11 of Z204-D
- Logic 1 at pin 5 of Z210
- Logic 1 at pin 5 of Z211
- Logic 1 at pin 5 of Z212

Chapter 7

Operator's Control Panel

Operating Modes 1 and 2 give the human operator some manual control over the machine via a hand-held control panel and an umbilical cord. The schematic in Fig. 7-1 shows that the Operator's Control Panel is electrically simple—nothing more than nine pushbutton switches.

The control panel is hardly sophisticated enough to perform the tasks required for a suitable manual control scheme. The system must include provisions for converting the signal from a single pushbutton into the corresponding 4-bit motor control code. Depressing the FORWARD LEFT pushbutton, for example, must generate FWD LEFT code, 0010.

Furthermore, the operator's control scheme should make it possible to initiate a desired motion by momentarily depressing one of the pushbutton switches. In other words, there should be no need to hold down a pushbutton in order to maintain a selected motion. This requires some data-latching circuitry.

So the simplicity of the Operator's Control Panel belies the sophistication of the circuitry that supports the scheme. That circuitry shown in Fig. 7-2, is located on the Main Electronics Board.

CIRCUIT DESCRIPTION

The switches on the operator's control panel are connected through a 10-conductor umbilical cable to J200 on the Main Electronics Board. Resistors R206 through R214 pull up the input con-

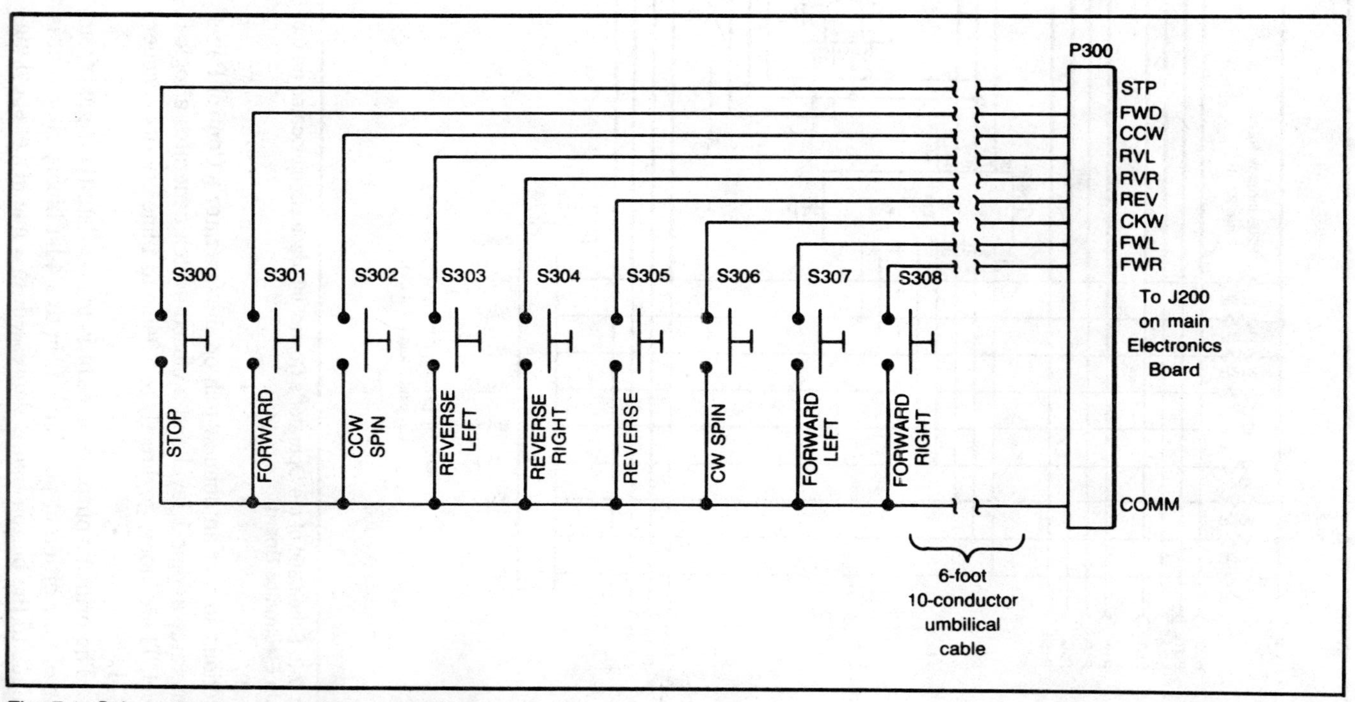

Fig. 7-1. Schematic diagram of the operator's control panel.

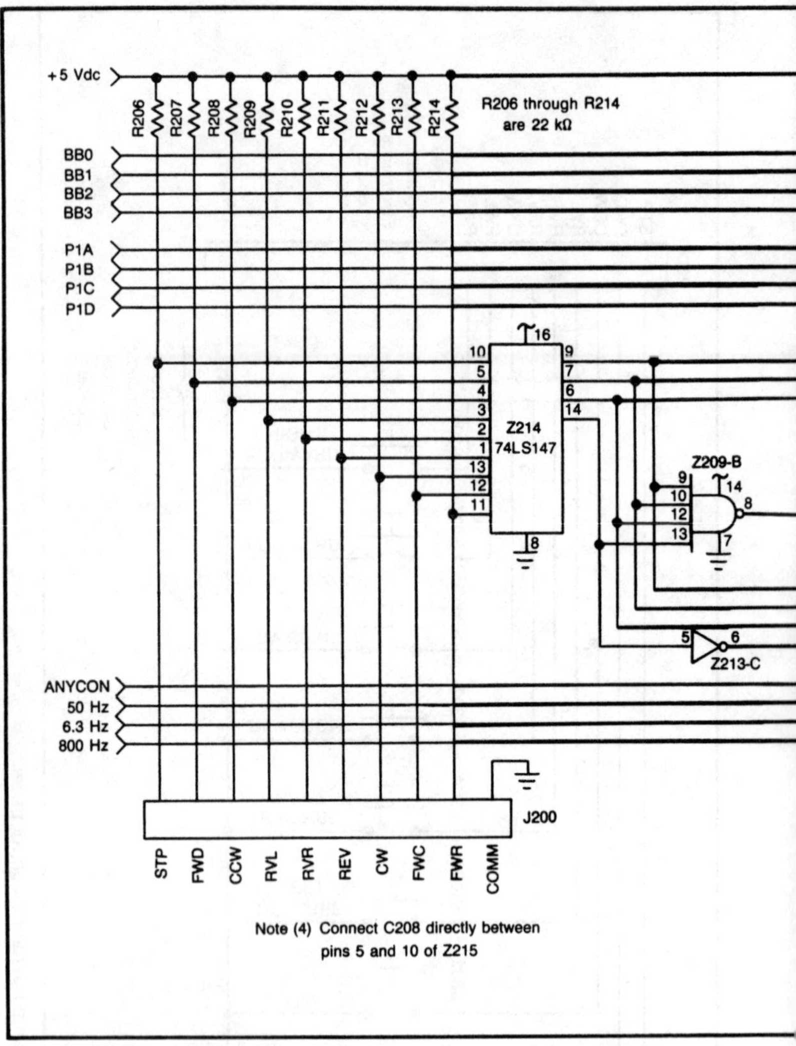

Fig. 7-2. Schematic of the Operator's Control Interface circuit located on the Main Electronics Board.

nections so that an open switch on the Operator's Control Panel generates a logic-1 level and a closed switch generates a logic-0 level. These inputs go directly to a 9-line to 4-line priority encoder, Z214.

The primary purpose of Z214 is to translate the signal from one of nine different pushbuttons in to a 4-bit binary code that is unique to the button that is depressed. In a manner of speaking,

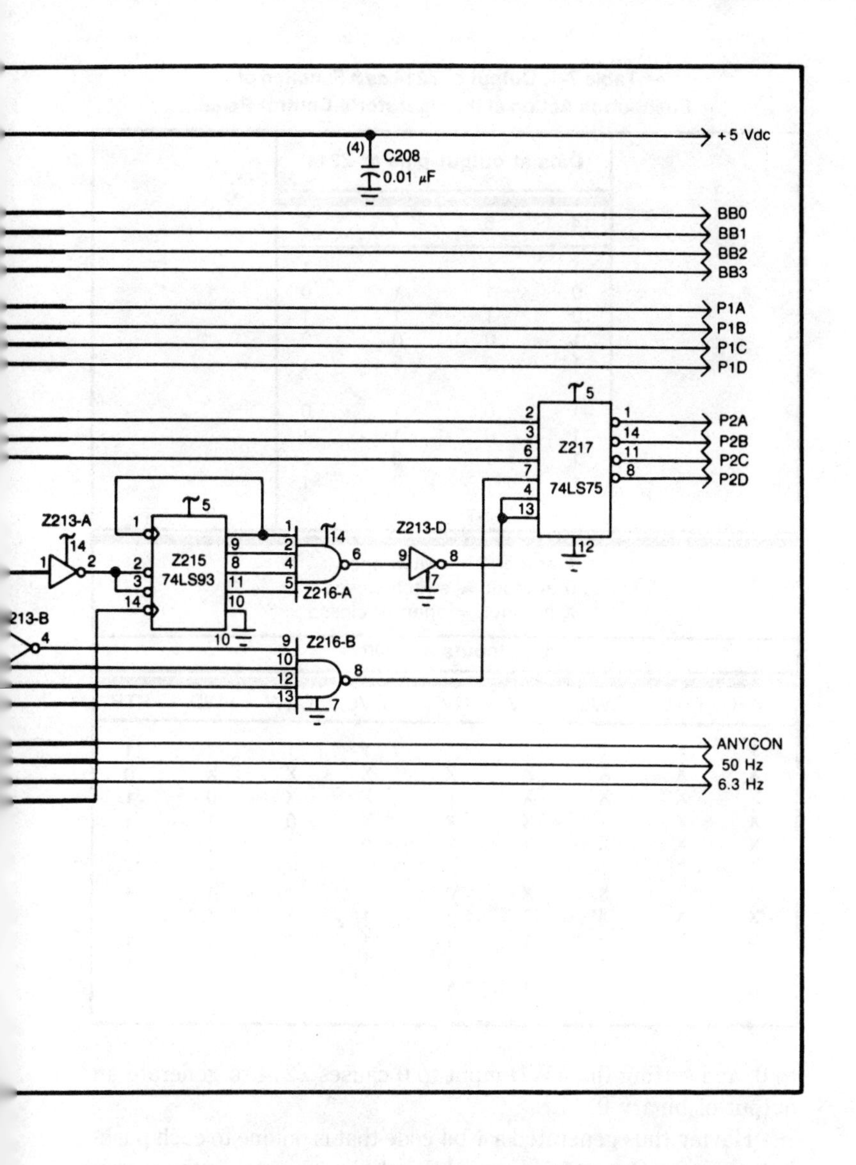

it serves the purpose of a decimal-to-binary converter. Table 7-1 shows the 4-bit binary codes that appear at the outputs of Z214 as a function of the button that is depressed at the Operator's Control Panel.

When no pushbuttons are depressed, for example, all inputs are at logic 1 and Z214 outputs binary 1111 on pins 14, 6, 7, and 9, respectively. Depressing the FORWARD pushbutton sets FWD

Table 7-1. Output of Z214 as a Function of Pushbutton Action at the Operator's Control Panel.

Inputs at J200									Data at output pins of Z214			
FWR	FWL	CW	REV	RVR	RVL	CCW	FWD	STP	14	6	7	9
1	1	1	1	1	1	1	1	1	1	1	1	1
X	X	X	X	X	X	X	X	0	0	1	1	0
X	X	X	X	X	X	X	0	1	0	1	1	1
X	X	X	X	X	X	0	1	1	1	0	0	0
X	X	X	X	X	0	1	1	1	1	0	0	1
X	X	X	X	0	1	1	1	1	1	0	1	0
X	X	X	0	1	1	1	1	1	1	0	1	1
X	X	0	1	1	1	1	1	1	1	1	0	0
X	0	1	1	1	1	1	1	1	1	1	0	1
0	1	1	1	1	1	1	1	1	1	1	1	0

1 at input = switch open
0 at input = switch closed
X at input = open or closed

to 0, and setting the FWD input to 0 causes Z214 to generate an output of binary 0111.

Having thus generated a 4-bit code that is unique to each pushbutton on the Operator's Control Panel, the next step is to convert these codes into the standard motor control codes. Depressing the FORWARD pushbutton, for instance, should eventually direct 0000—the forward motion code—to the D/S motor system. Table 7-2 compares the codes from Z214 with the standard motor codes that should appear later in the system as P2A, P2B, P2C, and P2D.

Obviously some codes conversion is necessary. Data in the Z214 data column must be converted to that shown in the motor response

Table 7-2. A Comparison of the Outputs from Z214 with the Motor Control Codes.

Switch Depressed and Desired Motor Response	Data at Z214 pins				Motor Response code			
	14	6	7	9	P2D	P2C	P2B	P2A
none	1	1	1	1				
STOP	0	1	1	0	1	0	0	1
FORWARD	0	1	1	1	0	0	0	0
CCW SPIN	1	0	0	0	0	1	1	1
REVERSE LEFT	1	0	0	1	0	1	1	0
REVERSE RIGHT	1	0	1	0	0	1	0	1
REVERSE	1	0	1	1	0	1	0	0
CW SPIN	1	1	0	0	0	0	1	1
FORWARD LEFT	1	1	0	1	0	0	1	0
FORWARD RIGHT	1	1	1	0	0	0	0	1

column. The conversion isn't as tricky as it might at first appear. With the exception of one single bit, the data in the motor response column is simply an inverted version of that in the Z214 data column. Most of the conversion can be done by simply inverting the data from Z214.

Figure 7-2 shows the three lower-order bits from Z214 connected directly to the corresponding inputs of a data latch, Z217. Outputs P2A, P2B, and P2C are taken from the latch's inverting outputs, so the conversion for these bits is complete at that point.

The only part of the conversion that cannot be accomplished by a simple inverting operation is the high-order bit in the FORWARD code. The output of Z214 is 0111, and the desired output code is 0000. The conversion is handled by invertors Z213-B and Z213-C and NAND gate Z216-B.

The purpose of the remaining circuitry in Fig. 7-2 is to debounce the switches on the Operator's Control Panel. The technique is the same one described earlier for the blunder input interface scheme. The general idea is to hold a binary counter at 0000 until a switch is closed. The counter is then allowed to run; and when the count reaches binary 1111, the new data is read into a latch. The delay is on the order of 20 milliseconds—plenty of time to let the switch-bounce effects settle down.

The data from the output of Z214 is binary 1111 only while

no pushbutton is being depressed. A 4-input NAND gate, Z209-B, generates a logic-0 output only as long as this condition exists. Z213-A inverts the no-switch level to logic 1 and applies it to the reset pins on a binary counter, Z215. So as long as no switch is depressed on the Operator's Control Panel, counter Z215 is stopped and cleared to binary 0000.

Depressing any switch on the Operator's Control Panel causes one or more zeroes to appear in the code from Z214. As a result, the output of Z209-B goes to logic 1, the output of Z213-A goes to logic 0, and the counter begins running from 0000. The clocking rate is fixed by the 800 Hz clock waveform applied to pin 14 of the counter.

About 20 milliseconds after the counting starts, the output of the counter reaches binary 1111. Z216-A responds by generating a logic-0 pulse. Z213-D inverts it to a logic-1, or positive-going, pulse. That pulse is applied to the gating inputs of the data latch—pins 4 and 13 of Z217. The output of the latch is thus updated.

The delay masks any contact-bounce effect from the Operator's Control Panel, and the latch feature retains the designated code at the P2 outputs when the operator releases the pushbutton.

CONSTRUCTION AND TEST HINTS FOR THE CONTROL PANEL

Table 7-3 is the parts list for the Operator's Control Panel that is shown schematically in Fig. 7-1, and Fig. 7-3 is a guide for mounting and labeling the switches.

Use Fig. 7-3 as a guide for mounting and labeling the pushbutton switches on the Operator's Control Panel. Bundle ten, 6-foot lengths of stranded wire, then run one end through a hole in the top end of the project box, and fix the cable in place with a suitable strain relief. Solder the switch connections as shown in the schematic.

Solder the loose ends of the conductors to plug assembly P300.

Table 7-3. Parts List for the Operator's Control Panel.

S300,S301, S302, S303, S304, S305, S306, S307, S308, S309—Normally open, momentary pushbutton switch
P300—15-pin, male D-subminiature plug
1 ea. Project box measuring approximately 2 3/4 × 4 × 2 inches.

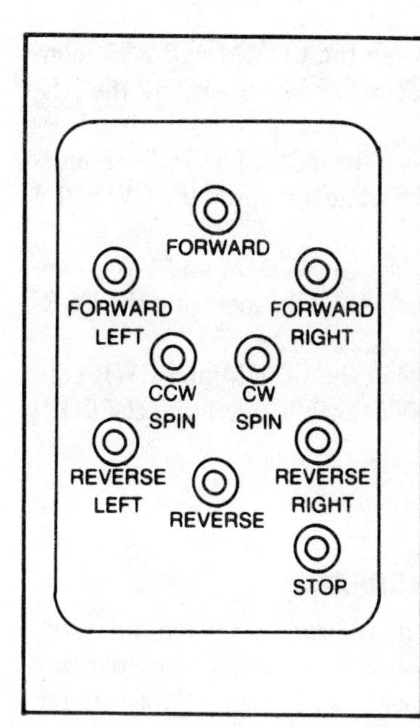

Fig. 7-3. Suggested layout and labeling scheme for the operator's control pane.

Use any convenient pin arrangement, but be sure to record each pin number and the function associated with it.

Use the following procedure to test the electrical integrity of all the connections. In each case, the meter should show zero resistance when the designated switch is depressed and infinite resistance when it is not.

- Connect an ohmmeter between the COMM and STP connections on P300 and note the response while operating the STOP switch.
- Connect an ohmmeter between the COMM and FWD connections on P300 and note the response while operating the FORWARD switch.
- Connect an ohmmeter between the COMM and CCW connections on P300 and note the response while operating the CCW SPIN switch.
- Connect an ohmmeter between the COMM and RVL connections on P300 and note the response while operating the REVERSE LEFT switch.
- Connect an ohmmeter between the COMM and RVR connections on P300 and note the response while operating the REVERSE RIGHT switch.

- Connect an ohmmeter between the COMM and REV connections on P300 and note the response while operating the REVERSE switch.
- Connect an ohmmeter between the COMM and CW connections on P300 and note the response while operating the CW SPIN switch.
- Connect an ohmmeter between the COMM and FWL connections on P300 and note the response while operating the FORWARD LEFT switch.
- Connect an ohmmeter between the COMM and FWR connections on P300 and note the response while operating the FORWARD RIGHT switch.

CONSTRUCTION AND TEST HINTS FOR THE CONTROL INTERFACE

Table 7-4 is the parts list for the Operator's Control Interface portion of the Main Electronics Board. Having those components at hand, use Fig. 7-4 as a guide for soldering resistors R206 through R214 to a 24-pin DIP header, CP201.

Make sure that the main power switch, S1, is turned OFF, then use Fig. 7-5 as a guide for mounting the following components to the Main Electronics Board:

- 24-pin DIP socket for CP201
- 14-pin DIP socket for Z213
- 16-pin DIP socket for Z214
- 14-pin DIP socket for Z215
- 14-pin DIP socket for Z216
- 16-pin DIP socket for Z217
- Connector J200

Use the schematic in Fig. 7-2 as a guide for making the necessary wire-wrap and solder connections. Make certain that the main power switch is turned OFF, then:

- Insert header CP201 into the designated socket.
- Insert ICs Z213 through Z217 into their sockets.
- Solder capacitor C208 directly across pins 5 and 10 if Z215.

Turn ON the main power switch and conduct the following tests to confirm the proper distribution of power to the IC devices:

Table 7-4. Parts list for the Operator's Control Interface on the Main Electronics Board.

R206, R207, R208, R209, R210, R211, R212, R213, R214—22 kΩ, 1/4 W resistor
Z213—74LS04 hex inverter
Z214—74LS145 or 74147 9-line to 4-line priority encoder
Z215—74LS93 4-bit binary counter
Z216—74LS40 dual 4-input NAND gate
Z217—74LS75 quad latch
J200—15-pin, PC-mount, female D-subminiature connector
3 ea. 14-pin DIP wire-wrap socket
2 ea. 16-pin DIP wire-wrap socket
1 ea. 24-pin DIP wire-wrap socket

- Measure about +5V dc at pin 14 of Z213.
- Measure about +5V dc at pin 16 of Z214.
- Measure about +5V dc at pin 5 of Z215.
- Measure about +5V dc at pin 14 of Z216.
- Measure about +5V dc at pin 5 of Z217.

Install the Operator's Control Panel by inserting P300 on the umbilical cable into J200 on the main electronics board, then conduct the following tests.

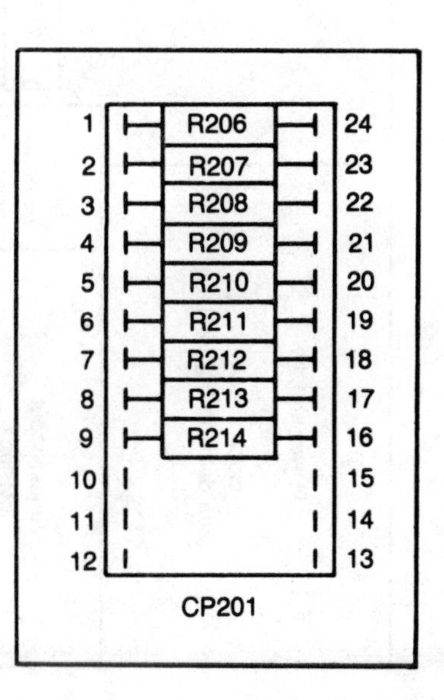

Fig. 7-4. Suggested arrangement of resistors R206 through R214 on the DIP header, CP201.

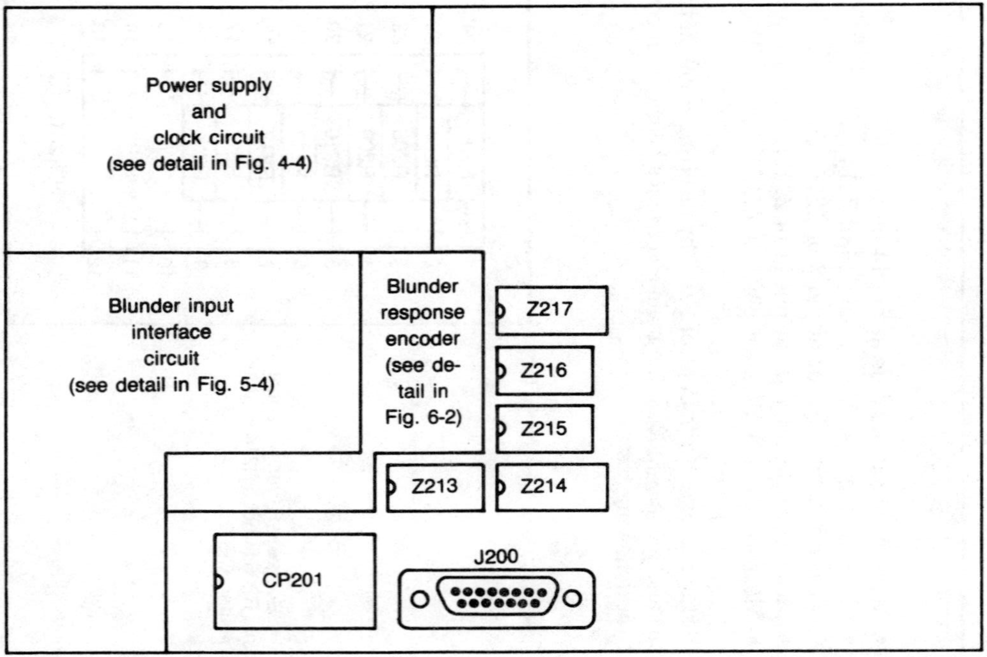

Power supply
and
clock circuit
(see detail in Fig. 4-4)

Blunder input
interface
circuit
(see detail in Fig. 5-4)

Blunder
response
encoder
(see de-
tail in
Fig. 6-2)

Z217

Z216

Z215

Z213

Z214

CP201

J200

Fig. 7-5. Suggested layout of Operator's Control Interface Portion of the Main Electronics Board.

Make certain that no switches are depressed on the Operator's Control Panel and note:

- Logic 1 at the STP connection J200
- Logic 1 at the FWD connection J200
- Logic 1 at the CCW connection J200
- Logic 1 at the RVL connection J200
- Logic 1 at the RVR connection J200
- Logic 1 at the REV connection J200
- Logic 1 at the CW connection J200
- Logic 1 at the FWL connection J200
- Logic 1 at the FWR connection J200
- Logic 1 at pin 10 of Z214
- Logic 1 at pin 5 of Z214
- Logic 1 at pin 4 of Z214
- Logic 1 at pin 3 of Z214
- Logic 1 at pin 2 of Z214
- Logic 1 at pin 1 of Z214
- Logic 1 at pin 13 of Z214
- Logic 1 at pin 12 of Z214
- Logic 1 at pin 11 of Z214
- Logic 1 at pin 2 of Z217
- Logic 1 at pin 3 of Z217
- Logic 1 at pin 6 of Z217
- Logic 0 at pin 9 of Z216-B
- Logic 0 at pin 13 of Z216-B
- Logic 1 at pin 7 of Z217
- Logic 0 at pin 8 of Z209-B
- Logic 1 at pins 2 and 3 of Z215
- Logic 1 at pin 6 of Z216-A
- Logic 0 at pins 4 and 13 of Z217
- 800 Hz clock waveform at pin 14 of Z215

Hold down the STOP pushbutton on the operator's control panel and note the following:

- Logic 0 at pin 10 of Z214
- Logic 0 at pin 9 of Z214
- Logic 0 at pin 14 of Z214
- Logic 0 at pin 7 of Z217
- Logic 1 at pin 8 of Z209-B
- Logic 0 at pins 2 and 3 of Z215
- 50 Hz positive-going pulses at pins 4 and 13 of Z217

Depress and then release the STOP pushbutton on the Operator's Control Panel. With no pushbuttons depressed, note:

- Logic 1 at pin 1 of Z217
- Logic 0 at pin 14 of Z217
- Logic 0 at pin 11 of Z217
- Logic 1 at pin 8 of Z217

Hold down the FORWARD pushbutton and note the

- Logic 0 at pin 5 of Z214.

Release the FORWARD pushbutton and note:

- Logic 0 at pin 1 of Z217
- Logic 0 at pin 14 of Z217
- Logic 0 at pin 11 of Z217
- Logic 0 at pin 8 of Z217

Hold down the CCW SPIN pushbutton and note

- Logic 0 at pin 4 of Z214.

Release the CCW SPIN pushbutton and note:

- Logic 1 at pin 1 of Z217
- Logic 1 at pin 14 of Z217
- Logic 1 at pin 11 of Z217
- Logic 0 at pin 8 of Z217

Hold down the REVERSE LEFT pushbutton and note

- Logic 0 at pin 3 of Z214.

Release the REVERSE LEFT pushbutton and note:

- Logic 0 at pin 1 of Z217
- Logic 1 at pin 14 of Z217
- Logic 1 at pin 11 of Z217
- Logic 0 at pin 8 of Z217

Hold down the REVERSE RIGHT pushbutton and note

- Logic 0 at pin 2 of Z214.

Release the REVERSE RIGHT pushbutton and note:

- Logic 1 at pin 1 of Z217
- Logic 0 at pin 14 of Z217
- Logic 1 at pin 11 of Z217
- Logic 0 at pin 8 of Z217

Hold down the REVERSE pushbutton and note

- Logic 0 at pin 1 of Z214.

Release the REVERSE pushbutton and note:

- Logic 0 at pin 1 of Z217
- Logic 0 at pin 14 of Z217
- Logic 1 at pin 11 of Z217
- Logic 0 at pin 8 of Z217

Hold down the CW SPIN pushbutton and note

- Logic 0 at pin 13 of Z214.

Release the CW SPIN pushbutton and note:

- Logic 1 at pin 1 of Z217
- Logic 1 at pin 14 of Z217
- Logic 0 at pin 11 of Z217
- Logic 0 at pin 8 of Z217

Hold down the FORWARD LEFT pushbutton and note

- Logic 0 at pin of Z214.

Release the FORWARD LEFT pushbutton and note:

- Logic 0 at pin 1 of Z217
- Logic 1 at pin 14 of Z217
- Logic 0 at pin 11 of Z217
- Logic 0 at pin 8 of Z217

Hold down the FORWARD RIGHT pushbutton and note

- Logic 0 at pin 11 of Z214.

Release the FORWARD RIGHT pushbutton and note:

- Logic 1 at pin 1 of Z217
- Logic 0 at pin 14 of Z217
- Logic 0 at pin 11 of Z217
- Logic 0 at pin 8 of Z217

Chapter 8

System Mode Control

The Mode Control circuit described in this chapter is the one responsible for coordinating all of Buster's activities. It accepts motion-control data from four different sources, but directs just one of those sources to the motor control bus at any given time. It oversees the timing operations for the blunder sequences and it reads the Mode Control selector switches. The Mode Control circuit is clearly the traffic controller for the entire Buster system.

The exact nature of the Mode Control operations is determined by the setting of a pair of mode-select switches. As shown in Table 8-1, there are four different operating modes that are labeled 0 through 3.

Notice that all but Mode 1 use both a *default* operating phase and an *interrupt* operating phase. The default phases represent the activity that is normally in effect: standing motionless, accepting motion codes from the operator's control panel, or moving forward across the floor. The interrupt phase represents the activity that takes place when a blunder condition occurs. The only interrupt activity built into this Buster system is the blunder response cycle.

Mode 0 is a standby mode. Power consumption is at a minimum and the machine appears to be turned off. The blunder response system is switched on, however, so Buster is able to respond instantly to any contact with the blunder switches.

Mode 1 gives the human operator full control over Buster's motion via the Operator's Control Panel and umbilical cord. The blun-

Table 8-1. Summary of Buster's Operating Modes.

Operating Mode 0	
Mode switch settings:	K1 = 0 K0 = 0
Default motion code:	STOP
Interrupt condition:	Blunder bumper contact
Return-to-default condition:	No blunder contact for about 3 consecutive seconds
General description:	Buster remains motionless until a blunder bumper is touched
Operating Mode 1	
Mode switch settings:	K1 = 0 K0 = 1
Default motion code:	Codes from the operator's control panel
Interrupt condition:	None
Return-to-default condition:	None
General description:	All motion control commands are taken from the operator's control panel
Operating Mode 2	
Mode switch settings:	K1 = 1 K0 = 0
Default motion code:	Codes from the operator's control panel
Interrupt condition:	Blunder bumper contact
Return-to-default condition:	No blunder contact for about 3 consecutive seconds
General description:	Motion codes are taken from the Operator's Control panel except during the execution of blunder responses
Operating Mode 3	
Mode switch settings:	K1 = 1 K0 = 1
Default motion code:	FORWARD
Interrupt condition:	Blunder bumper contact
Return-to-default condition:	No blunder contact for about 3 consecutive seconds
General description:	Buster runs forward except during the execution of blunder responses

der feature is disabled, so it is up to the operator to maneuver Buster through contact situations in the environment.

Mode 2 also gives the human operator control over Buster's motion, but only while the system is not executing a blunder sequence. When a blunder condition occurs, Buster overrides any commands from the control panel and takes care of the situation in his own fashion. Control returns to the operator when the blunder condition is cleared.

Mode 3 gives Buster the opportunity to roam about on his own. The default motion code, FORWARD, suggests that he normally runs straightforward. Bumper contacts interrupt that activity and initiate a blunder response sequence. Without a doubt, it is the most exciting operating mode.

CIRCUIT DESCRIPTION

Figure 8-1 is a block diagram of the mode select circuitry. First notice that there are four sources of data, or ports, that feed information to the motor control bus. Port 0 provides a STOP motion code, Port 1 represents motion codes from the blunder response circuitry, Port 2 provides motion codes from the Operator's Control Panel, and Port 3 always generates a FWD (forward) motion code. It is possible to generate all of Buster's mode-control activity from combinations of those four ports.

All four ports make their respective 4-bit motion codes available at all times, only one should be connected to the motor control bus at any moment. This is the purpose of the port select circuit.

The port select circuit continuously monitors the status of Mode Select switches K0 and K1, the ANYCON signal from the blunder input interface, and the two clock signals (50 Hz and 6.3 Hz) from the power supply and clock circuits. This small family of signals is all the information that the port select circuit needs to coordinate all of Buster's activities and ensure that only one port is providing data to the motor control bus.

The diagram shows port-select connections going to Ports 1 through 3. Output 1 from the port select circuit enables Port 1, output 2 enables Port 2, and output 3 enables Port 3. But what about Port 0?

Port 0 is taken as the default port. In other words, Port 0 places its STOP motion code onto the motor control bus whenever all three of the other ports are disabled.

Figure 8-2 is the schematic diagram for the system mode control circuit. The main output from the circuit—the motor control bus—is made up of the four lines, MB0 through MB3. This bus terminates outputs from buffers Z222 and Z223 with pull-up resistors R217 through R220.

Z222 is a noninverting, 3-state buffer circuit that handles two sets of 4-bit data independently. Whenever pin 1 of that device is pulled down to logic 0, bits P1A through P1D are directed to the motor control bus. In other words, setting pin 1 of Z222 to logic 0 sends blunder response codes to the motor control system.

Setting pin 19 of Z222 to logic 0 enables the second half of the device and delivers bits P2A through P2B to the motor control bus. This action sends signals from the Operator's Control Panel to the motor control system. So Z222 handles the tasks of Port 1 and 2, with pin 1 enabling Port 1, and pin 19 enabling Port 2.

Z223 is an inverting version of the port-control buffer circuit.

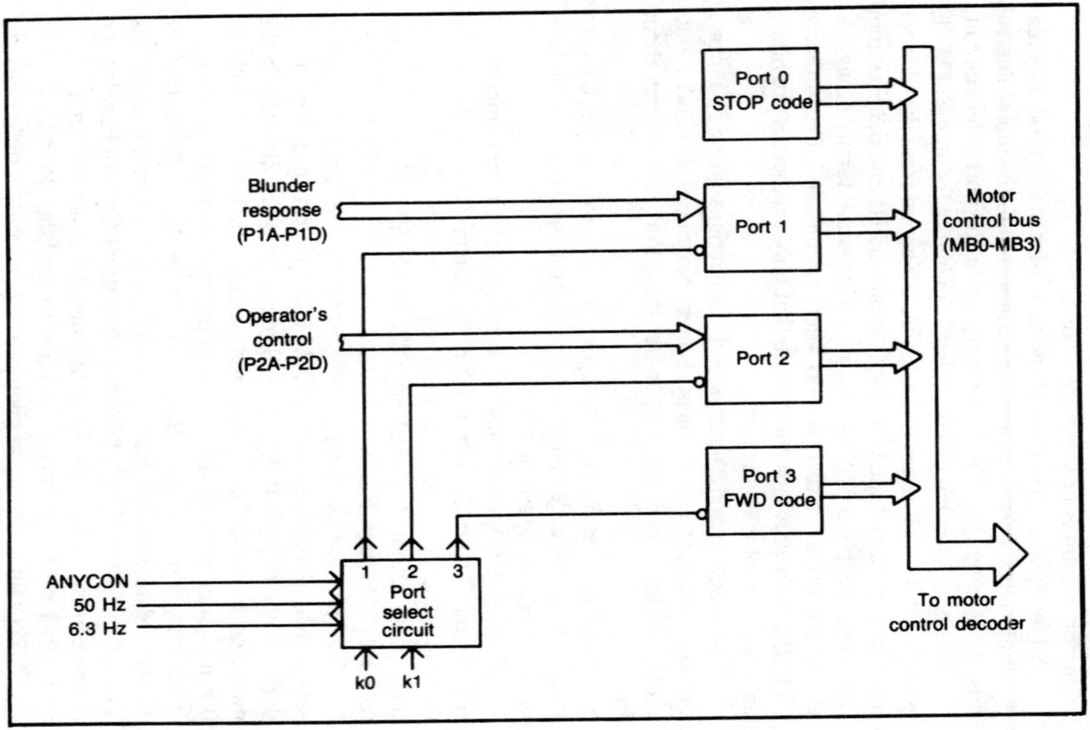

Fig. 8-1. Block diagram of the Mode Select scheme.

Pulling pin 1 to logic 0 enables half the device and sends inverted versions of the logic levels at pins 2, 4, 6, and 8 to the motor control bus. These four inputs are fixed at logic 1; so when the port is enabled, it sends binary 0000—the FWD motion code—to the motor control system.

Half of Z223 handles the task of Port 3, and it sends the FWD motion code to the motor control bus whenever pin 1 is set to logic 0. The other half of the device serves a different purpose that is described in the next chapter.

Resistors R217 through R220 are connected directly between the motor control bus and + 5 Vdc and tend to pull all four bus connections to logic 1. Unless Port 1, 2, or 3 is enabled, the pull-up resistors define the Port-0 default operation by sending binary 1111—a STOP code—to the motor system.

Table 8-2 summarizes the operation of the port circuits. Note that Ports 1, 2, and 3 are enabled by setting their respective enabling pins to logic 0. Port 0 is enabled whenever the other three parts are disabled.

Much of the remaining circuitry is devoted to port select operations. The two outputs from Z221 select Ports 1 and 2, while the output from NAND gate Z218-B selects Port 3. It is difficult to get a real appreciation for the operation of the port select circuit, without first understanding how it times the blunder sequence.

Recall that ANYCON is a logic level that indicates whether or not any of the blunder contact switches are closed. ANYCON has a logic-0 level whenever no blunder switches are closed, and it takes on logic-1 level as long as one or more switches are closed.

Buster should begin a blunder response sequence the moment any sort of blunder condition occurs—when ANYCON changes from logic 0 to logic 1. The blunder sequence should not end, until two conditions are met: ANYCON returns to logic 0 and a timer circuit subsequently times its full cycle.

The ANYCON signal is important to the function of the port select circuitry. The onset of the ANYCON condition starts the blunder cycle, and the end of the ANYCON condition initiates the time-delay operation.

The situation invites the application of a flip-flop circuit. The onset of the ANYCON signal sets the flip-flop and the end of the delay operation resets it. For the sake of this discussion, the output of the flip-flop is called the BSEQ (Blunder SEQuence) signal. BSEQ is set to logic 1 at the onset of a blunder condition, and it remains at logic 1 until the blunder condition is cleared and the sub-

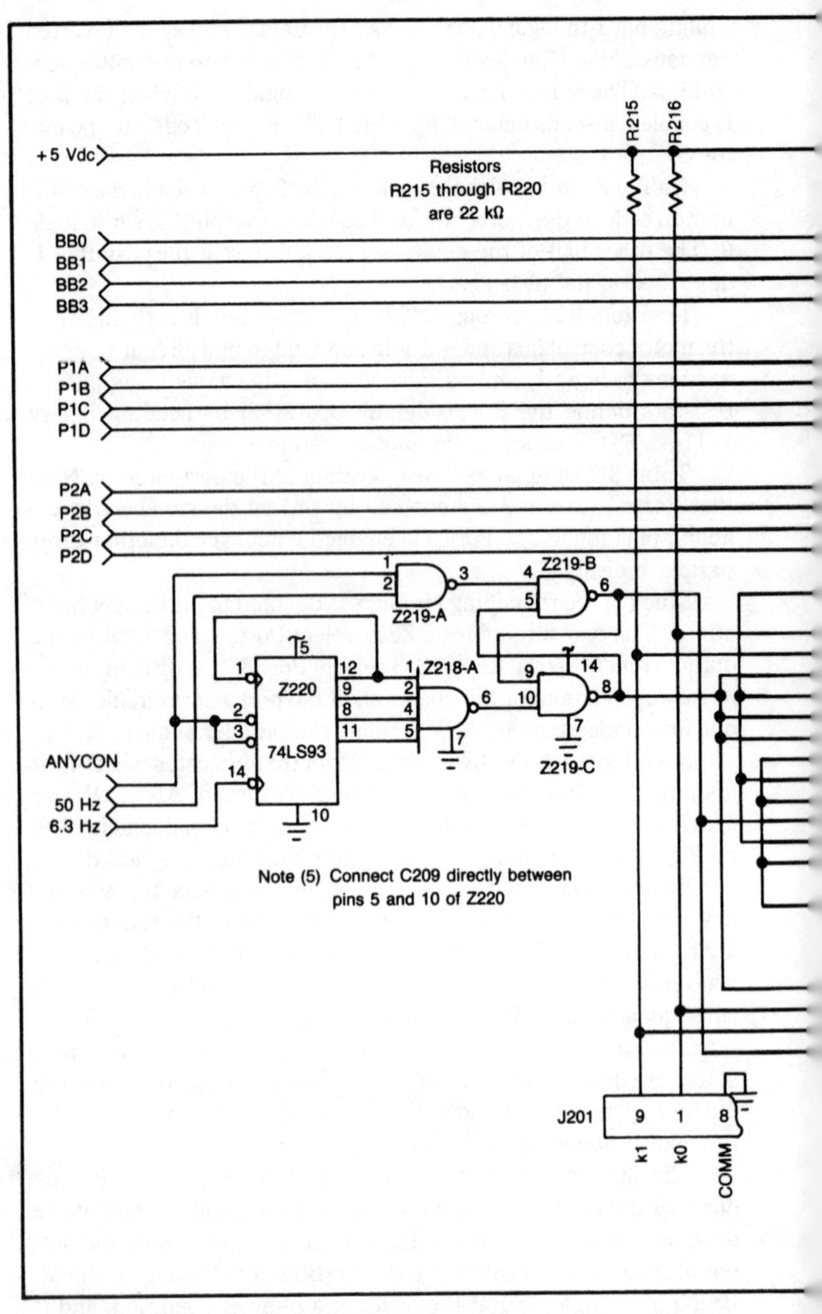

Fig. 8-2. Schematic diagram of the Mode Select portion of the Main Electronics Board.

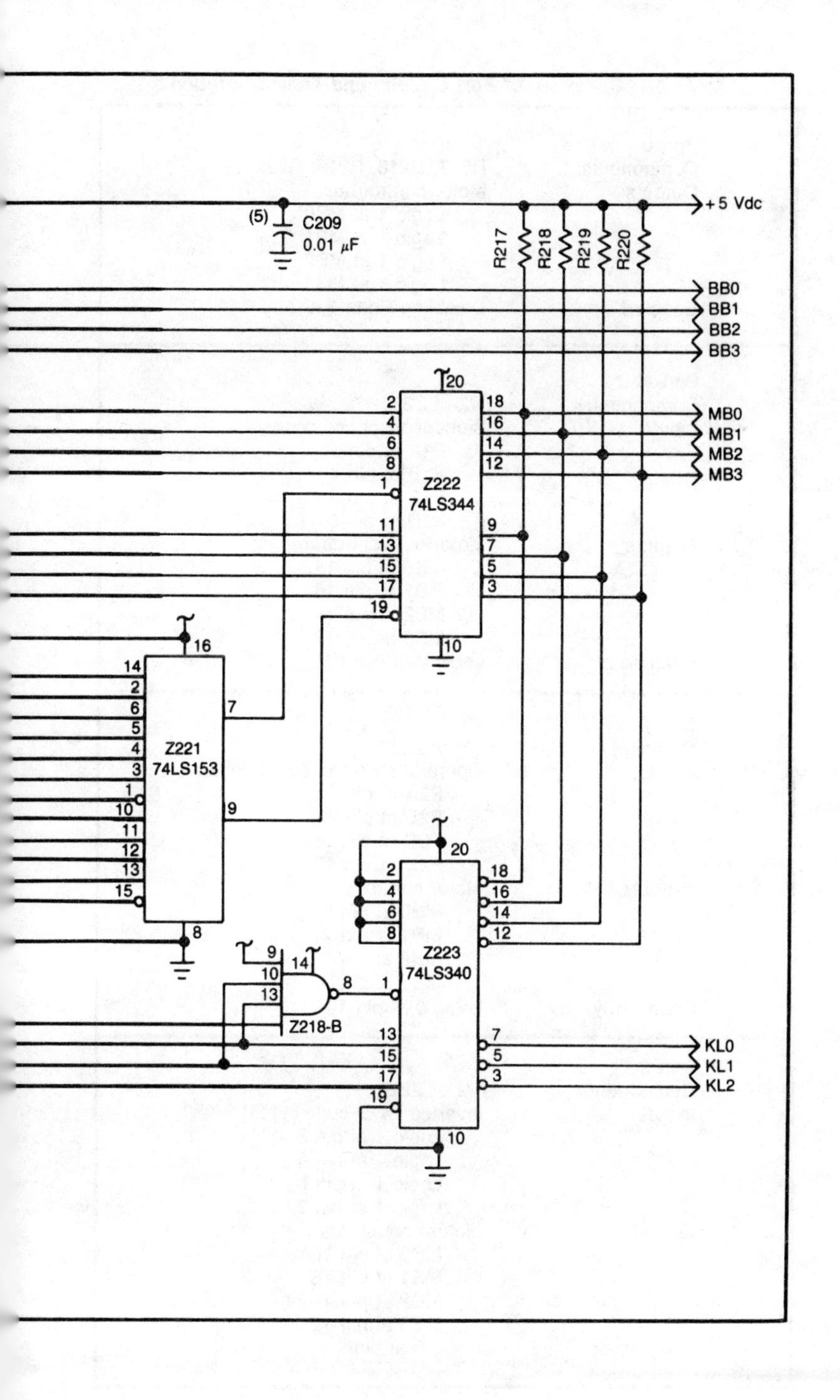

+5 Vdc
(5)
C209
0.01 µF
R217
R218
R219
R220
BB0
BB1
BB2
BB3
MB0
MB1
MB2
MB3
Z222
74LS344
Z221
74LS153
Z223
74LS340
Z218-B
KL0
KL1
KL2

Table 8-2. Summary of Port Circuits and Their Operation.

Port 0	
Components:	R217, R218, R219, R220
Outputs:	Motor control bus
	Logic 1 at MB0
	Logic 1 at MB1
	Logic 1 at MB2
	Logic 1 at MB3
Enabled by:	Disabling Ports 1, 2 and 3
Port 1	
Component:	1/2 of Z222
Inputs:	Blunder response codes
	P1A at pin 2
	P1B at pin 4
	P1C at pin 6
	P1D at pin 8
Outputs:	Motor control bus
	MB0 at pin 18
	MB1 at pin 16
	MB2 at pin 14
	MB3 at pin 12
Enabled by:	Logic 0 at pin 1
Port 2	
Component:	1/2 of Z222
Inputs:	Operator's control panel codes
	P2A at pin 11
	P2B at pin 13
	P2C at pin 15
	P2D at pin 17
Outputs:	Motor control bus
	MB0 at pin 9
	MB1 at pin 7
	MB2 at pin 5
	MB3 at pin 3
Enables by:	Logic 0 at pin 19
Port 3	
Component:	1/2 of Z223
Inputs:	Inverted FWD code (1111)
	Logic 1 at pin 2
	Logic 1 at pin 4
	Logic 1 at pin 6
	Logic 1 at pin 8
Outputs:	Motor control bus
	MB0 at pin 18
	MB1 at pin 16
	MB2 at pin 14
	MB3 at pin 12
Enabled by:	Logic 0 at pin 1

Table 8-3. Summary of Port-Select Operations.

Mode	BSEQ	Port	Operating Phase
0	0	0	STOP code to motor system
0	1	1	Blunder response to motor system
1	X	2	Operator's control to motor system
2	0	2	Operator's control to motor system
2	1	1	Blunder response to motor system
3	0	3	FWD code to motor system
3	1	1	Blunder response to motor system

X = irrelevant

sequent time delay is completed.

In short, Buster executes a blunder response as long as BSEQ is at the logic-1 level. Table 8-3 illustrates how BSEQ affects the port-select procedure within each of the four operating modes.

When operating in Mode 0, for example, the system selects Port 0 as long as BSEQ is at logic 0—as long as Buster is not executing a blunder sequence. But the system switches to Port 1 when a blunder contact occurs, and BSEQ is subsequently set to logic 1.

BSEQ is not relevant in Mode 1 because the human operator has total control over the system at all times. It is relevant in Mode 2, however. Working in Mode 2, the system uses the Operator's Control instructions through Port 2 as long as BSEQ is at logic 0. But when a blunder contact occurs, BSEQ switches to logic 1 and the system switches Port 1 to the motor control bus.

Mode 3 uses Port 3 to place the FWD motion code onto the motor control bus at all times except when BSEQ is at logic 1. Through this blunder sequence, the system switches the blunder codes onto the motor control bus through Port 1.

The blunder flip-flop appears in Fig. 8-2 in the form of two NAND gates, Z219-B and Z219-C. Pin 6 of Z219-B is the BSEQ terminal, and pin 8 of Z219-C is an inverted version of the same signal.

The BSEQ signal is set by applying negative-going pulses to pin 4 of Z219-B. This pin is connected to the output of another NAND gate, Z219-A, which takes its inputs from ANYCON and the 50 Hz clock source. The flip-flop and Z219-A thus work together to set and maintain the BSEQ sequence as long as ANYCON is at logic 1.

The BSEQ signal is cleared by applying a negative-going pulse to pin 10 of Z291-C. This happens whenever the output of NAND gate Z218-A senses a count of binary 1111 from counter Z220. Z220, however, is stopped and held at binary 0000 as long as the ANY-CON signal at pins 2 and 3 is at logic 1. In other words, the counter cannot run as long as a blunder condition exists.

The moment the blunder condition is cleared and ANYCON drops to logic 0, counter Z220 beings clocking at a rate determined by the frequency of the clock signal applied to pin 14. If the circuit uses the 6.3 Hz clock signal as suggested by the schematic diagram, the time required to count from binary 0000 to binary 1111 is on the order of 2.6 seconds. Provided another blunder contact doesn't interrupt the timing operation, the binary 1111 output from thc counter circuit ultimately resets the BSEQ flip-flop to conclude the blunder sequence.

Table 8-1 shows how the Mode Control switches, K0 and K1, are used to set Buster's current operating mode. Table 8-2 shows the pin locations and logic levels required for enabling the ports, and Table 8-3 shows how the BSEQ logic level should affect the port-selection operation. Table 8-4 assembles all of the information into a port-select encoder table.

The first two lines in the table show the encoding that is neces-

Table 8-4. Function Encoder Table for the Mode Select Circuit.

K1	K0	BSEQ	PE1	PE2	PE3
0	0	0	1	1	1
0	0	1	0	1	1
0	1	0	1	0	1
0	1	1	1	0	1
1	0	0	1	0	1
1	0	1	0	1	1
1	1	0	1	1	0
1	1	1	0	1	1

PE1 = logic level applied to the Port-1 enabling terminal, pin 1 of Z222

PE2 = logic level applied to the Port-2 enabling terminal, pin 19 of Z222

PE3 = logic level applied to the Port-3 enabling terminal, pin 1 of Z223

Table 8-5. Parts Lists for the Mode Select Circuits. (A) The Mode Switch and Lamp Assembly. (B) The Mode Control Portion of the Main Electronics Board.

(A)

L400, L401, L402, L403, L404, L405, L406, L407, L408, L409, L410—Any common red LED

S400, S401—Subminiature SPST toggle switch

J400—16-pin wire-wrap DIP socket

1 length of 16-conductor ribbon cable with 16-pin DIP plugs on both ends

(B)

R215, R216, R217, R218, R219, R220—22 kΩ, 1/4W resistor

C209—0.01 μF capacitor

Z218—74LS40 dual 4-input NAND gate
Z219—74LS00 quad 2-input NAND gate
Z220—74LS93 4-bit binary counter
Z221—74SLS153 dual 4-line to 1-line data multiplexer
Z222—74SL344 8-bit, 3-state noninverting buffer
Z223—74LS340 8-bit, 3-state inverting buffer

J201—16-pin DIP socket
CP202—24-pin DIP header
CJ202—24-pin wire-wrap DIP socket

3 ea. 14-pin wire-wrap DIP socket
1 ea. 16-pin wire-wrap DIP socket
2 ea. 20-pin wire-wrap DIP socket

sary when operating in Mode 0—when K0 and K1 are both set to logic 0. As long as BSEQ is at 0, none of the active ports is enabled, and the system resorts to using the binary 1111 motion code as determined by the pull-up resistors connected to the motor control bus. But when BSEQ is at logic 1, the table shows that Port 1 should be enabled. In this case, a logic-0 level is applied to pin 1 of Z222, and the blunder response codes, P1A through P1D, are applied to the motor control bus. All of the encoding is handled by the pattern of connections to pins 5, 4, 3, and 1 of Z221.

CONSTRUCTION HINTS

Table 8-5 is the parts lists for the Mode Switch and Lamp Panel assembly and the Mode Control portion of the Main Electronics Board.

Use Fig. 8-3 as a guide for cutting a panel and laying out the components for the Mode Select and Lamp Panel. Mount the two toggle switches in such a way that flipping them upward sets them to their OFF position.

Make the electrical connections as shown in the schematic diagram in Fig. 8-4.

Mount the Mode Switch and Lamp Panel at a readily accessible location on the main platform assembly, making sure that your 16-conductor jumper cable will reach between the panel and J201 on the Main Electronics Board.

Solder resistors R215 through R220 to the 24-pin DIP header, CP202, as shown in Fig. 8-5.

Use Fig. 8-6 as a guide for mounting the following components to the Main Electronics Board:

- 16-pin wire-wrap DIP socket for J201
- 24-pin wire-wrap DIP socket, CJ202
- 20-pin wire-wrap DIP socket for Z223
- 20-pin wire-wrap DIP socket for Z222
- 16-pin wire-wrap DIP socket for Z221
- 14-pin wire-wrap DIP socket for Z218
- 14-pin wire-wrap DIP socket for Z219
- 14-pin wire-wrap DIP socket for Z220

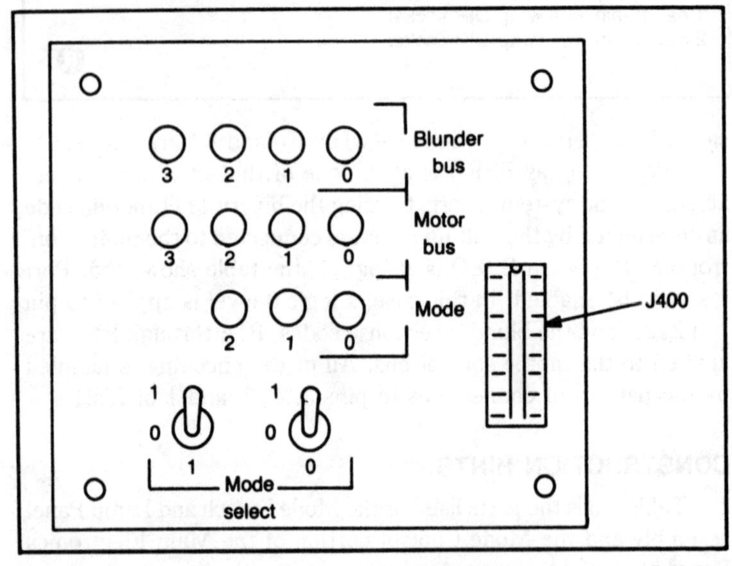

Fig. 8-3. Suggested physical layout of the Mode Switch and Lamp Panel.

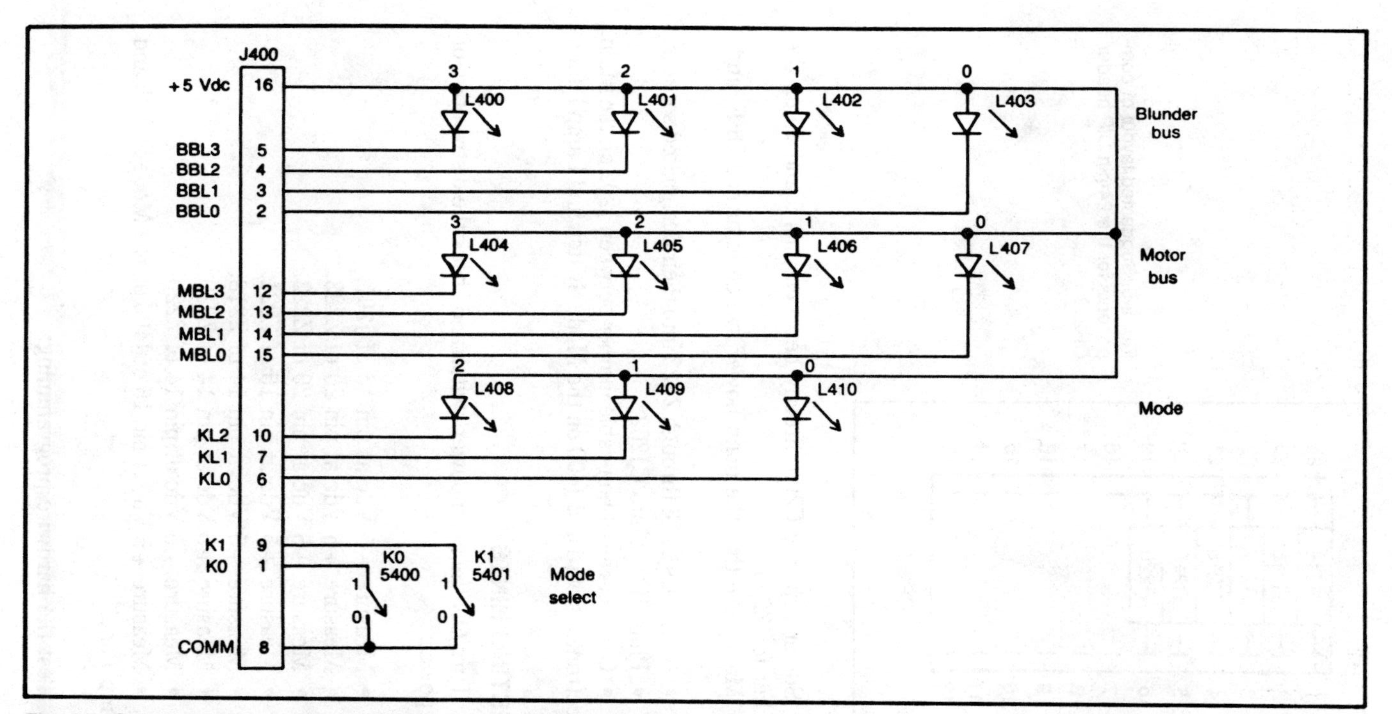

Fig. 8-4. Schematic diagram of the Mode Switch and Lamp Panel.

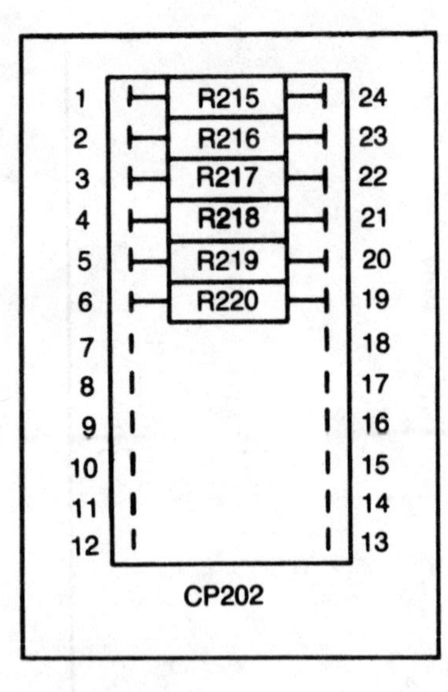

Fig. 8-5. Suggested layout of components for the 24-pin DIP header, CP202.

Solder capacity C209 directly between pins 5 and 10 on the socket for Z220.

Make sure that the main power switch is turned OFF, and:

- Insert ICs Z218 through Z223 into their designated sockets.
- Plug CP202 into CJ202.
- Connect the ribbon-cable jumper between J202 on the Main Electronics Board and J400 on the Mode Switch and Lamp Panel.

TESTING HINTS

Turn ON the main power switch and check the distribution of dc power:

- Measure +5 Vdc at pin 16 of J201.
- Measure +5 Vdc at pin 20 of Z223.
- Measure +5 Vdc at pin 20 of Z222.
- Measure +5 Vdc at pin 16 of Z221.
- Measure +5 Vdc at pin 14 of Z218.
- Measure +5 Vdc at pin 14 of Z219.
- Measure +5 Vdc at pin 14 of Z220.
- Measure +5 Vdc at pin 16 of J400 on the Mode Switch and Lamp Panel.

Test the hardwire programming:

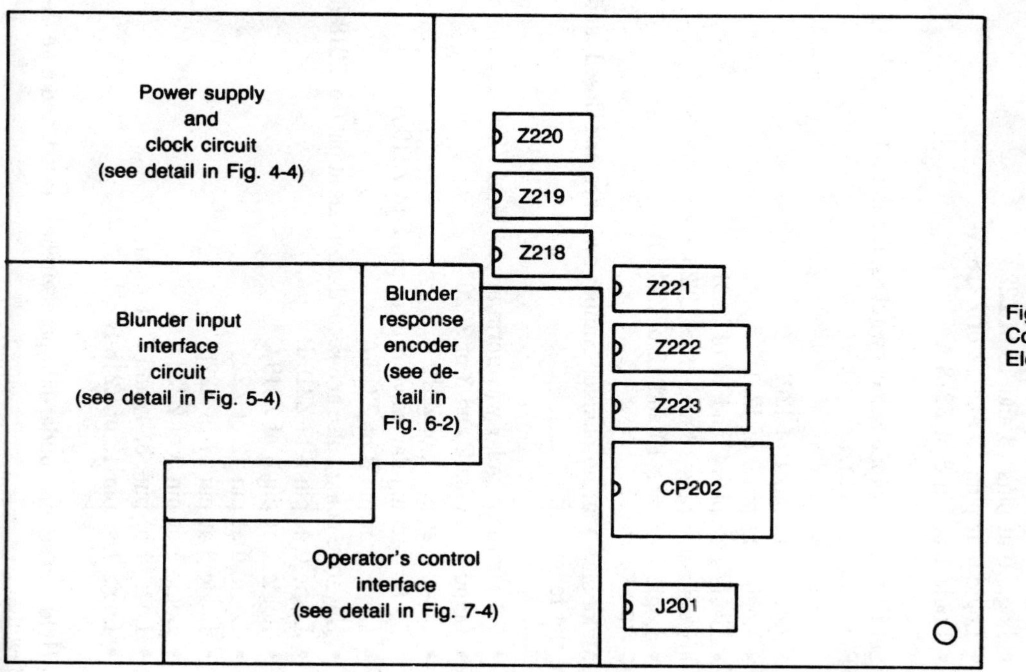

Fig. 8-6. Suggested layout of Mode Control components on the Main Electronics Board.

- Logic 1 at pins 5, 12 and 13 of Z221
- Logic 0 at pin 11 of Z221
- Logic 1 at pins 2, 4, 6, and 8 of Z223

Set both of the Mode Select switches, K0 and K1, to their "0" (ON) positions, and note:

- Logic 0 at pin 1 of J201
- Logic 0 at pin 9 of J201
- Logic 0 at pins 2 and 14 of Z221
- Logic 0 at pins 13 and 15 of Z223
- Logic 0 at pins 10 and 12 of Z218-A
- Logic 1 at pin 8 of Z218-A

Set both of the Mode Select switches, K0 and K1, to their "1" (OFF) positions, and note:

- Logic 1 at pin 1 of J201
- Logic 1 at pin 9 of J201
- Logic 1 at pins 2 and 14 of Z221
- Logic 1 at pins 13 and 15 of Z223
- Logic 1 at pins 10 and 12 of Z218-A

Make sure that no blunder-bumper switches are closed, and note the following:

- 6.3 Hz rectangular waveform at pin 14 of Z220
- Logic 0 at pins 2 and 3 of Z220
- Logic 0 at pin 1 of Z219-A
- 50 Hz rectangular waveform at pin 2 of Z219-A
- Logic 1 at pin 3 of Z219-A
- Negative-going pulse at 3-second intervals at pin 6 of Z218-A
- Logic 0 at pin 6 of Z219-B
- Logic 1 at pin 8 of Z219-C
- Logic 0 at pin 12 of Z221
- Logic 0 at pin 17 of Z223
- Logic 1 at pin 3 of Z223
- Logic 1 at pins 3, 4, and 6 of Z221
- Logic 1 at pin 13 of Z218-B

Hold any set of blunder-bumper switches closed, and note the following:

- Logic 1 at pins 2 and 3 of Z220
- Logic 1 at pin 1 of Z219-A
- 50 Hz rectangular waveform at pin 3 of Z219-A
- Logic 1 at 6 of Z218-A
- Logic 1 at pin 6 of Z219-B
- Logic 0 at pin 8 of Z219-C
- Logic 1 at pin 12 of Z221
- Logic 1 at pin 17 of Z223
- Logic 0 at pin 3 of Z223
- Logic 0 at pins 3, 4, and 6 of Z221
- Logic 0 at pin 13 of Z218-B
- Logic 1 at pin 8 of Z218-B
- Logic 1 at pin 1 of Z223

Set the Mode Switches for Mode-0 operation (K0 = 0, K1 = 0), make sure no blunder switches are closed, and note the following:

- Logic 1 at pin 1 of Z222
- Logic 1 at pin 19 of Z222
- Logic 1 at pin 1 of Z223
- Logic 1 at pin 7 of Z223
- Logic 1 at pin 5 of Z223
- Logic 1 at pin 3 of Z223

With the Mode Switches still set for Mode-0 operation as above, hold closed any set of blunder-bumper switches, and note the following:

- Logic 0 at pin 1 of Z222
- Logic 1 at pin 19 of Z222
- Logic 1 at pin 1 of Z223
- Logic 1 at pin 7 of Z223
- Logic 1 at pin 5 of Z223
- Logic 0 at pin 3 of Z223

Set the Mode Switches for Mode-1 operation (K0 = 1, K1 = 0), make sure no blunder switches are closed, and note the following:

- Logic 1 at pin 1 of Z222
- Logic 0 at pin 19 of Z222
- Logic 1 at pin 1 of Z223
- Logic 0 at pin 7 of Z223

- Logic 1 at pin 5 of Z223
- Logic 1 at pin 3 of Z223

With the Mode Switches still set for Mode-1 operation as above, hold closed any set of blunder-bumper switches, and note:

- Logic 1 at pin 1 of Z222
- Logic 0 at pin 19 of Z222
- Logic 1 at pin 1 of Z223
- Logic 0 at pin 7 of Z223
- Logic 1 at pin 5 of Z223
- Logic 0 at pin 3 of Z223

Set the Mode Switches for Mode-2 operation (K0-0, K1-1), make sure no blunder switches are closed, and note the following:

- Logic 1 at pin 1 of Z222
- Logic 0 at pin 19 of Z222
- Logic 1 at pin 1 of Z223
- Logic 1 at pin 7 of Z223
- Logic 0 at pin 5 of Z223
- Logic 1 at pin 3 of Z223

With the Mode Switches still set for Mode-2 operation as above, hold closed any set of blunder-bumper switches, and note:

- Logic 0 at pin 1 of Z222
- Logic 1 at pin 19 of Z222
- Logic 1 at pin 1 of Z223
- Logic 1 at pin 7 of Z223
- Logic 0 at pin 5 of Z223
- Logic 0 at pin 3 of Z223

Set the Mode Switches for Mode-3 operation (K0 = 1, K1 = 1), make sure no blunder switches are closed, and note the following:

- Logic 1 at pin 1 of Z222
- Logic 1 at pin 19 of Z222
- Logic 0 at pin 1 of Z223
- Logic 0 at pin 7 of Z223
- Logic 0 at pin 5 of Z223
- Logic 1 at pin 3 of Z223

With the Mode Switches still set for Mode-3 operation as above, hold closed any set of blunder-bumper switches, and note:

- Logic 0 at pin 1 of Z222
- Logic 1 at pin 19 of Z222
- Logic 1 at pin 1 of Z223
- Logic 0 at pin 7 of Z223
- Logic 0 at pin 5 of Z223
- Logic 0 at pin 3 of Z223

Chapter 9

Final Bus Encoder Circuits

The Final Bus Encoder circuits described in this chapter complete the work on the Main Electronics Board. One part of the circuit simply interfaces the bus circuits with the lamps on the Mode Switch and Lamp Panel. A more critical part of the circuit transforms motor bus codes into forms that suit the requirement of the D/S Motor Output Board.

CIRCUIT DESCRIPTION

Figure 9-1 is the schematic diagram of the Final Bus Encoder circuit. The inputs to the circuit are:

1. The 4-bit blunder bus (BB0 through BB3)
2. The 4-bit motor bus (MB0 through MB3)
3. The 3-bit mode bus (KL0 through KL3)

The blunder- and motor-bus codes are delivered to this circuit in their raw, TTL form. The purpose of Z227 is to invert and amplify the current level of these bus signals, making them suitable for driving the LEDs on the Mode Switch and Lamp Panel. The mode-bus signals are already in this form by the time they reach this portion of the circuit.

The motor-bus signals are also applied to NAND gate Z219-D and multiplexers Z224 and Z225. These devices convert the motor-bus codes to the versions required for operating the D/S Motor Output Panel. Z226 inverts the codes and boosts the current level to make them compatible with the electrical requirements of the Motor Output Panel. Table 9-1 summarizes these encoding operations.

The Motor Bus Data, bits MB0 through MB3, represent the current motor code. The source of the data depends on Buster's current operating mode. The output of the circuit—the Motor Control Data—feeds information directly to the Motor Output Board.

CONSTRUCTION HINTS

Table 9-2 is the parts list for the Final Bus Encoder circuits, and Fig. 9-2 is a suggested layout of the Final Bus Encoder components on the Main Electronics Board. Figure 9-3 is a suggested layout for the resistors on the 24-pin DIP header, CP203.

Solder the resistors to CP203 and make the wire-wrap connections to the IC sockets and J201. Make the solderless connections to barrier strip TS202, but do not connect TS202 to the D/S Motor Output Board at this time.

Make sure the main power switch on the Power Distribution Panel is turned OFF, then insert the ICs into their sockets. If you have not already done so, connect the 16-conductor ribbon cable between J201 on the Main Electronics Board and J400 on the Mode Switch and Lamp Panel.

SUGGESTED TEST PROCEDURES

Turn ON the main power switch and note the distribution of power:

- +5 Vdc at pin 16 of Z224
- +5 Vdc at pin 16 of Z225
- +5 Vdc at pin 14 of Z226
- +5 Vdc at pin 20 of Z227
- 0 V at pins 1 and 19 of Z227
- +5 Vdc at pin 12 of Z219-D
- +5 Vdc at pin 16 of J201
- +5 Vdc at pin 16 of T400 on the Mode Switch and Lamp Panel

Make sure that no blunder switches are closed and note the following:

- Logic 0 at pin 2 of Z227

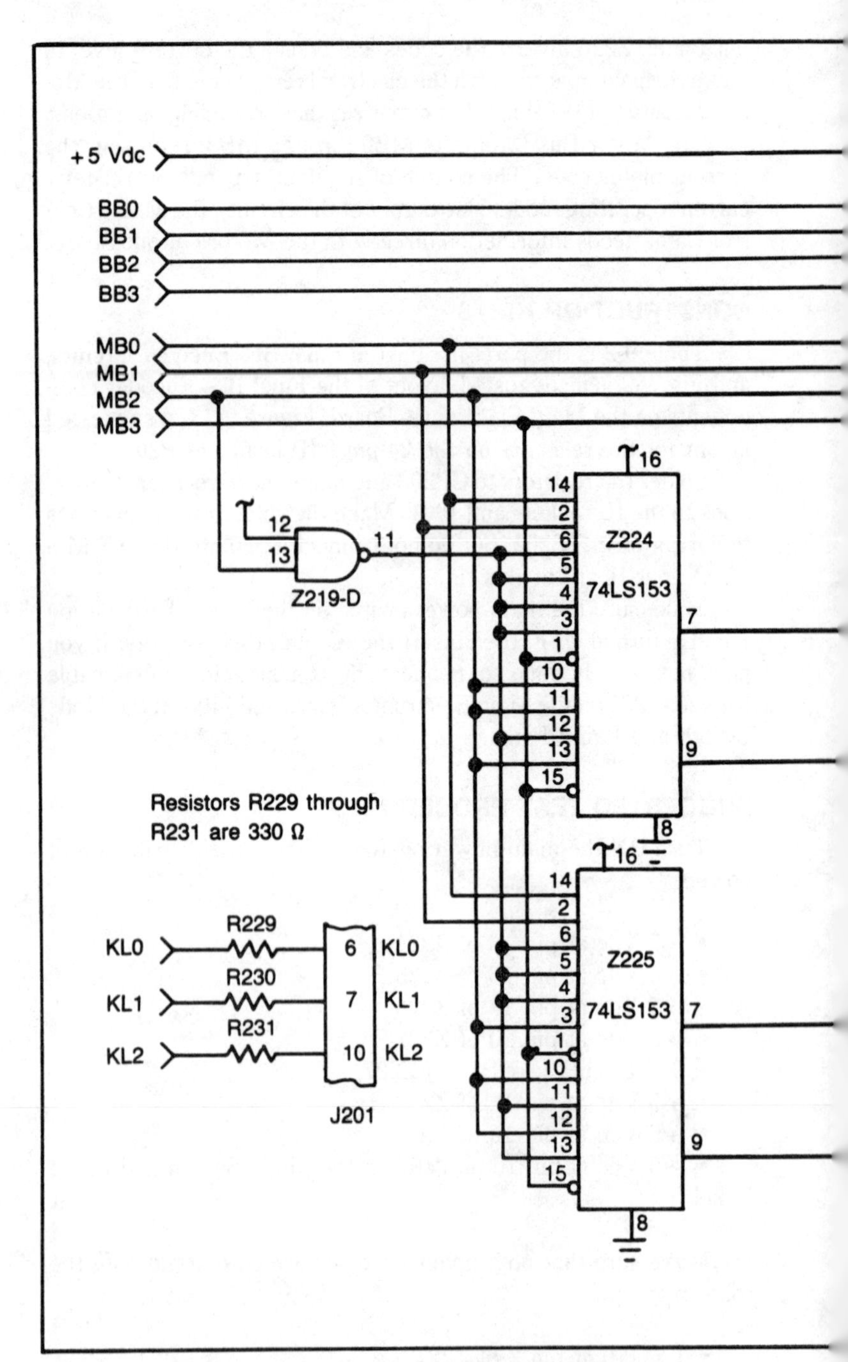

Fig. 9-1. Schematic diagram of the Final Bus Encoder circuits.

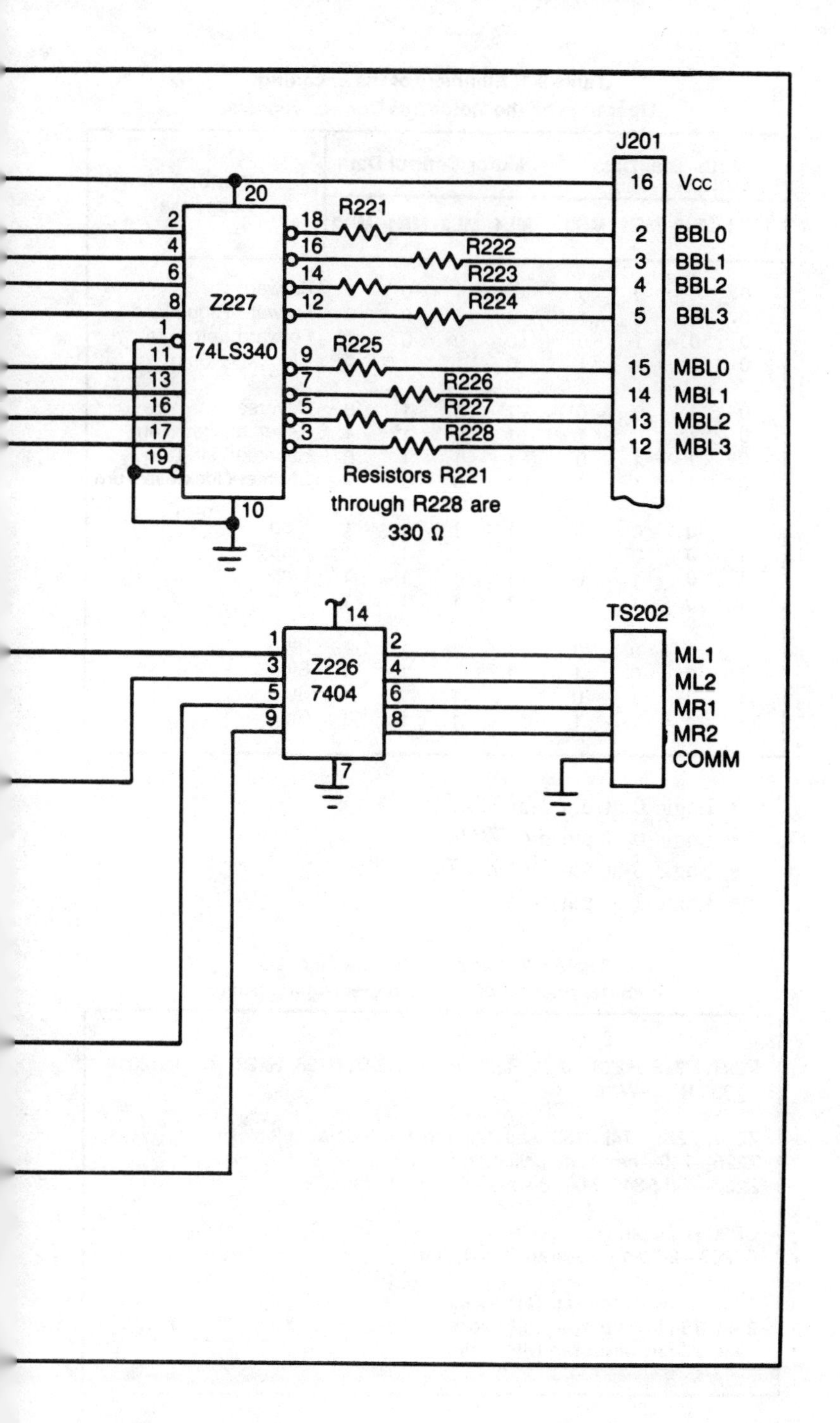
J201
16 Vcc
2 BBL0
3 BBL1
4 BBL2
5 BBL3
15 MBL0
14 MBL1
13 MBL2
12 MBL3
Z227
74LS340
R221
R222
R223
R224
R225
R226
R227
R228
Resistors R221 through R228 are 330 Ω
Z226
7404
TS202
ML1
ML2
MR1
MR2
COMM

Table 9-1. Summary of the Encoding Operation on the Motor-Bus Control Signals.

Motor Bus Data				Motor Control Data				Platform Response
MB3	MB2	MB1	MB0	ML1	ML2	MR1	MR2	
0	0	0	0	0	1	0	1	Forward
0	0	0	1	0	1	0	0	Forward Right Turn
0	0	1	0	0	0	0	1	Forward Left Turn
0	0	1	1	0	1	1	0	Clockwise Spin
0	1	0	0	1	0	1	0	Reverse
0	1	0	1	1	0	1	1	Reverse Right Turn
0	1	1	0	1	1	1	0	Reverse Left Turn
0	1	1	1	1	0	0	1	Counter-clockwise Turn
1	0	0	0	1	1	1	1	Stop
1	0	0	1	1	1	1	1	Stop
1	0	1	0	1	1	1	1	Stop
1	0	1	1	1	1	1	1	Stop
1	1	0	0	1	1	1	1	Stop
1	1	0	1	1	1	1	1	Stop
1	1	1	0	1	1	1	1	Stop
1	1	1	1	1	1	1	1	Stop

- Logic 0 at pin 4 of Z227
- Logic 0 at pin 6 of Z227
- Logic 0 at pin 8 of Z227
- Logic 1 at pin 18 of Z227

Table 9-2. Parts List for the Final Bus Encoder Section of the Main Electronics Board.

R221, R222, R223, R224, R225, R226, R227, R228, R229, R230, R231—330 Ω, 1/4W resistor

Z224, Z225—74LS153 dual 4-line to 1-line data multiplexer
Z226—740—hex open-collector inverter
Z227—74LS340 8-bit, 3-state inverting buffer

CP203—24-pin DIP header
CJ203—24-pin wire-wrap DIP socket

1 ea. 14-pin wire-wrap DIP socket
2 ea. 16-pin wire-wrap DIP socket
1 ea. 20-pin wire-wrap DIP socket

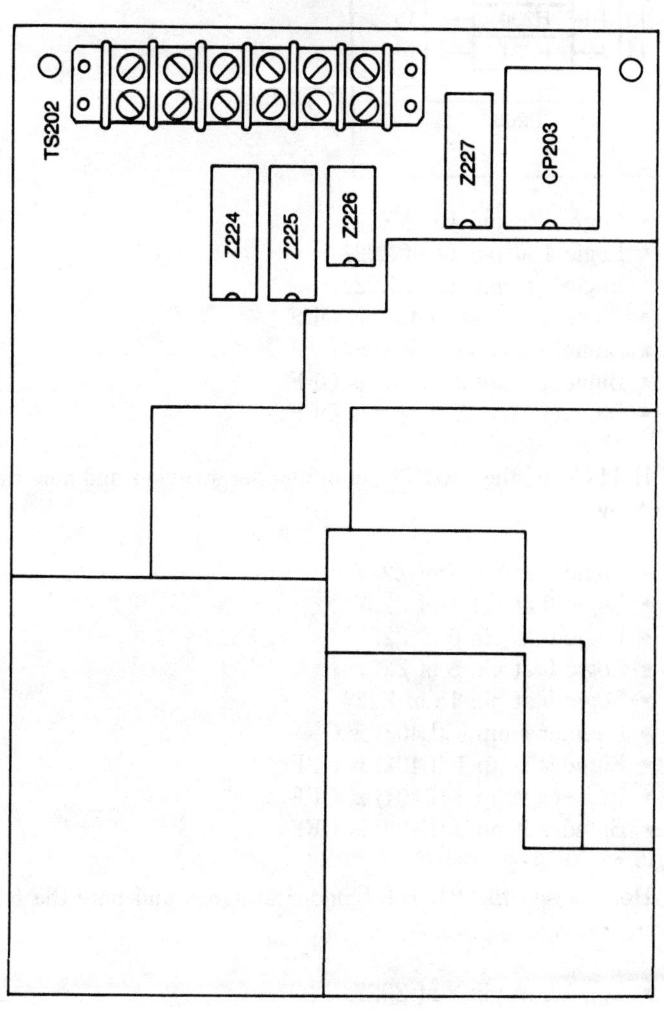

Fig. 9-2. Suggested layout of the Final Bus Encoder components on the Main Electronics Board.

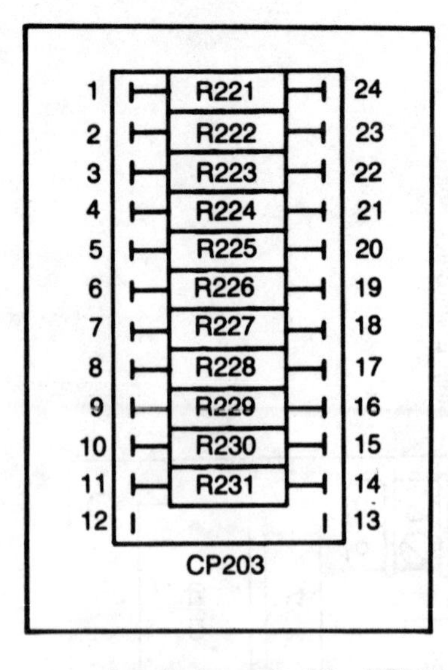

Fig. 9-3. Suggested layout of resistors on CP203.

- Logic 1 at pin 16 of Z227
- Logic 1 at pin 14 of Z227
- Logic 1 at pin 12 of Z227
- Blunder lamp 0 (L403) is OFF
- Blunder lamp 1 (L402) is OFF
- Blunder lamp 2 (L401) is OFF
- Blunder lamp 3 (L400) is OFF

Hold closed the FRONT set of blunder switches and note the following:

- Logic 1 at pin 2 of Z227
- Logic 0 at pin 4 of Z227
- Logic 0 at pin 6 of Z227
- Logic 0 at pin 8 of Z227
- Logic 0 at pin 18 of Z227
- Blunder lamp 0 (L403) is ON
- Blunder lamp 1 (L402) is OFF
- Blunder lamp 2 (L401) is OFF
- Blunder lamp 3 (L400) is OFF

Hold closed the RIGHT blunder switches and note the following:

- Logic 0 at pin 2 of Z227

- Logic 1 at pin 4 of Z227
- Logic 0 at pin 6 of Z227
- Logic 0 at pin 8 of Z227
- Logic 0 at pin 16 of Z227
- Blunder lamp 0 (L403) is OFF
- Blunder lamp 1 (L402) is ON
- Blunder lamp 2 (L401) is OFF
- Blunder lamp 3 (L400) is OFF

Hold closed the REAR blunder switches and note the following:

- Logic 0 at pin 2 of Z227
- Logic 0 at pin 4 of Z227
- Logic 1 at pin 6 of Z227
- Logic 0 at pin 8 of Z227
- Logic 0 at pin 14 of Z227
- Blunder lamp 0 (L403) is OFF
- Blunder lamp 1 (L402) is OFF
- Blunder lamp 2 (L401) is ON
- Blunder lamp 3 (L400) is OFF

Hold closed the LEFT blunder switches and note the following:

- Logic 0 at pin 2 of Z227
- Logic 0 at pin 4 of Z227
- Logic 0 at pin 6 of Z227
- Logic 1 at pin 8 of Z227
- Logic 0 at pin 12 of Z227
- Blunder lamp 0 (L403) is OFF
- Blunder lamp 1 (L402) is OFF
- Blunder lamp 2 (L401) is OFF
- Blunder lamp 3 (L400) is ON

Set the Mode Select switches for Mode-0 operation (K0 = 0, K1 = 0), make sure no blunder switches are closed, and note the following:

- Mode lamp 0 (L410) is OFF
- Mode lamp 1 (L409) is OFF
- Mode lamp 2 (L408) is OFF
- Mode bus lamp 0 (L407) is ON
- Motor bus lamp 1 (L406) is ON

- Motor bus lamp 2 (L405) is ON
- Motor bus lamp 3 (L404) is ON
- Logic 1 at pin 11 of Z227
- Logic 1 at pin 13 of Z227
- Logic 1 at pin 16 of Z227
- Logic 1 at pin 17 of Z227
- Logic 0 at pin 9 of Z227
- Logic 0 at pin 7 of Z227
- Logic 0 at pin 5 of Z227
- Logic 0 at pin 3 of Z227
- Logic 1 at pin 13 of Z219-D
- Logic 0 at pins 3, 4, 5, 6, and 12 of Z224
- Logic 1 at pins 2, 10, 11, 13, and 14 of Z224
- Logic 0 at pins 4, 5, 6, 11, and 13 of Z225
- Logic 1 at pins 2, 3, 10, 12, and 14 of Z225
- Logic 0 at pins 1, 3, 5, and 9 of Z226

Set the Mode-Select switches for Mode-1 operation (K0 = 1, K1 = 0) and note the following:

- Mode lamp 0 (L410) is ON.
- Mode lamp 1 (L409) is OFF.
- Mode lamp 2 (L408) is OFF.

Set the Mode Select switches for Mode-2 operation (K0 = 0, K1 = 1), make sure no blunder switches are closed, and note the following:

- Mode lamp 0 (L410) is OFF.
- Mode lamp 1 (L409) is ON.
- Mode lamp 2 (L408) is OFF.

Set the Mode Select switches for Mode-3 operation (K0 = 1, K1 = 1), make sure no blunder switches are closed, and note the following:

- Mode lamp 0 (L410) is ON
- Mode lamp 1 (L409) is ON
- Mode lamp 2 (L408) is OFF
- Motor bus lamp 0 (L407) is OFF
- Motor bus lamp 1 (L406) is OFF
- Motor bus lamp 2 (L405) is OFF

- Motor bus lamp 3 (L404) is OFF
- Logic 0 at pin 11 of Z227
- Logic 0 at pin 13 of Z227
- Logic 0 at pin 16 of Z227
- Logic 0 at pin 17 of Z227
- Logic 1 at pin 9 of Z227
- Logic 1 at pin 7 of Z227
- Logic 1 at pin 5 of Z227
- Logic 1 at pin 3 of Z227
- Logic 0 at pin 13 of Z219-D
- Logic 1 at pins 3, 4, 5, 6, and 12 of Z224
- Logic 0 at pins 2, 10, 11, 13, and 14 of Z224
- Logic 1 at pins 4, 5, 6, 11, and 13 of Z225
- Logic 0 at pins 2, 3, 10, 12, and 14 of Z225
- Logic 0 at pins 3 and 9 of Z226
- Logic 1 at pins 1 and 5 of Z226

Chapter 10

Finishing the Project

Wake up your family, call your friends; break out the root beer, ice, and potato chips! The job is nearly done, and it is time to celebrate.

Buster isn't a very sophisticated device in terms of modern robotics—even modern personal robotics. But it has been a big job, and your Buster doesn't have to take second place to any other robot.

MAKING THE FINAL CONNECTIONS

All that remains to be done is to make the five solderless connections between TS202 on the Main Electronics Board and TS102 on the Motor Output Board. Make sure the Main Power switch is turned OFF, then use AWG #18 stranded wire to make the following solderless connections:

- from COMM on TS120 to COMM on TS202
- from ML1 on TS120 to ML1 on TS202
- from ML2 on TS120 to ML2 on TS202
- from MR1 on TS120 to MR1 on TS202
- from MR2 on TS120 to MR2 on TS202

RUNNING THE FINAL TESTS

I strongly suggest that you run the final tests with Buster sitting on a table or workbench, supported in such a way that the drive wheels can turn without actually moving him. Also make sure the batteries have a good charge.

It is a good idea to set Buster to Mode 0, the sleep mode, before turning on the main power switch. So:

- Set the Mode Select switches for Mode 0.
- Turn ON the main power switch.
- Note that some of the indicator lamps turn on, but Buster does not move.
- Momentarily depress the FRONT blunder bumper, and note that the left motor turns in the reverse direction for about three seconds.
- Momentarily depress the RIGHT blunder bumper, and note that the both motors run in the forward direction for about three seconds.
- Momentarily depress the LEFT blunder bumper, and note that the left motor turns in the forward direction for about three seconds.

The operator's full-control mode, Mode 1, offers the best way to doublecheck Buster's responses to all possible commands. So make sure the Operator's Control is properly connected to the system, and set the Mode Select switches for Mode-1 operation. If Buster immediately responds by running one or both motors, depress the STOP pushbutton on the control panel.

The following series of Mode-1 tests call for momentarily depressing a pushbutton on the Operator's Control Panel and then indicating a motor response that should occur until you depress another pushbutton.

- Momentarily depress the FWD pushbutton and note that both motors run in the forward direction.
- Momentarily depress the CCW pushbutton and note that the left motor turns in the reverse direction and the right motor turns in the forward direction.
- Momentarily depress the RVL pushbutton and note that the right motor turns in the reverse direction.
- Momentarily depress the RVR pushbutton and note that the left motor turns in the reverse direction.
- Momentarily depress the REV pushbutton and note that both motors run in the reverse direction.

• Momentarily depress CW pushbutton and note that the right motor turns in the reverse direction and the left motor turns in the forward direction.

• Momentarily depress the FWL pushbutton and note that the right motor turns in the forward direction.

• Momentarily depress the FWR pushbutton and note that the left motor turns in the forward direction.

Assuming that Buster passes the tests just described, he is ready to begin running around the floor. So set him for the sleep mode—Mode 0—and place him on the floor. Momentarily depress the blunder bumpers as described earlier for Mode-0 tests, and note his 3-second responses.

Set him for Mode-1 operation, and use the Operator's Control Panel to move him around according to your commands.

Set him for Mode-2 operation and use the Operator's Control Panel to control him until he blunders into some sort of contract situation. Note how Buster works his way out of the contact situation and then resumes operation according to the last command you entered from the control panel.

When you are satisfied that Buster is in good working order, set him for Mode-3 operation. Now he is on his own.

Appendix A

Composite Buster Parts List

RESISTORS

330 Ω, 1/4 W resistor — 11 ea.

R221	R222	R223	R224
R225	R226	R227	R228
R229	R230	R231	

470 Ω, 1/4 W resistor — 4 ea.

R101	R102	R103	R104

1 KΩ, 1/4 W resistor — 4 ea.

R105	R106	R107	R108

1.8 KΩ, 1/4 W resistor — 1 ea.

15 KΩ, 1/4 W resistor — 1 ea.

R200

22 KΩ, 1/4 W resistor 19 ea.

R202	R203	R204	R205
R206	R207	R208	R209
R210	R211	R212	R213
R214	R215	R216	R217
R218	R219	R220	

CAPACITORS

0.01 μF capacitor 5 ea.

C203	C205	C206	C207
C209			

0.1 μF capacitor 3 ea.

C200	C201	C204

100 μF, 50 WVdc electrolytic capacitor 1 ea.

C202

MISCELLANEOUS SEMICONDUCTORS

2N2907A high-speed pnp transistor 4 ea.

Q101	Q102	Q103	Q104

1N4001 diode 4 ea.

D101	D102	D103	D104

Red LED 11 ea.

L400	L401	L402	L403
L404	L405	L406	L407
L408	L409	L410	

LM309K positive 5 V regulator 1 ea.

VR200

INTEGRATED CIRCUITS

LM555 timer 1 ea.

Z200

74LS00 quad 2-input NAND gate 2 ea.

Z204 Z219

74LS04 hex inverter 1 ea.

7406 hex open-collector inverter 1 ea.

Z226

74LS40 dual 4-input NAND gate 5 ea.

Z203 Z207 Z209 Z216
Z218

74LS75 quad latch 2 ea.

Z208 Z217

74LS93 four-bit binary counter 6 ea.

Z201 Z202 Z205 Z206
Z215 Z220

74LS147 or 74147 9-line to 4-line priority encoder 1 ea.

Z214

74LS151 8-line to 1-line data multiplexer 3 ea.

Z210 Z211 Z212

74LS153 dual 4-line to 1-line data multiplexer 3 ea.

Z221 Z224 Z225

74LS340 8-bit, 3-state inverting buffer 2 ea.

Z223 Z227

74LS344 8-bit, 3-state noninverting buffer 1 ea.

Z222

SWITCHES AND RELAYS

DPST toggle switch; 3 A minimum 1 ea.

S1

Lever-type microswitch 8 ea.
(such as Radio Shack #275-017) 8 ea.

S2 S3 S4 S5
S6 S7 S8 S9

Normally open, momentary pushbutton switch 10 ea.

S300 S301 S302 S303
S304 S305 S306 S307
S308 S309

Subminiature SPST toggle switch 2 ea.

S400 S401

SPDT relay — 4 ea.
minimum 1 A contact rating
6 Vdc, 1 mA coil

K101 K102 K103 K104

PLUGS AND CONNECTORS

Red binding post — 1 ea.
(such as Radio Shack #274-662)

J1

Black binding post — 1 ea.
(such as Radio Shack #274-662)

J2

PLUGS AND CONNECTORS

15-pin, male D-subminiature plug — 1 ea.

P300

15-pin, PC-mount, female D-subminiature connector — 1 ea.

J200

8-pin wire-wrap DIP socket — 1 ea.

14-pin wire-wrap DIP socket — 15 ea.

16-pin wire-wrap DIP socket — 11 ea.
(including J201, J400)

20-pin wire-wrap DIP socket — 3 ea.

24-pin wire-wrap DIP socket 4 ea.

CJ200 CJ201 CJ202 CJ203

24-pin DIP header

CP200 CP201 CP202 CP203

1 length of 16-conductor ribbon cable with 16-pin DIP plugs on both ends

BARRIER STRIPS

2-position, double-row barrier strip 1 ea.
(such as Radio Shack #274-656)

TS200

4-position, dual-row barrier strip 1 ea.
(such as Radio Shack #274-658)

TS1

6-position, dual-row barrier strip 6 ea.
(such as Radio Shack #274-659)

TS2 TS3 TS101 TS102
TS201 202

MOTORS AND BATTERIES

6 Vdc gearmotor wheel assembly 2 ea.

M1 M2

6 V motorcycle battery 2 ea.

B1 B2

FUSES

5 A slow-blow fuse 2 ea.

F1 F2

Fuseholder 2 ea.

MISCELLANEOUS COMPONENTS

TO-220 heat-sink assembly 1 ea.

Project box measuring approximately
2-3/4 × 4 × 2 inches 1 ea.

Appendix B

Main Electronics Board Details

Figure B-1 shows the suggested layout of parts for a finished version of the Main Electronics Board. Schematics and detailed instructions appear in Chapters 4 through 9.

The board, itself, is a 6 × 8 inch sheet of perfboard that has holes that are spaced for mounting IC sockets (1/10-inch apart). Use a small dab of vinyl-base adhesive—bathtub calk, for example—to fasten the wire-wrap IC sockets to the component side of the board.

Fashion a set of aluminum mounting brackets to secure the board vertically to the platform, just a couple of inches behind the batteries. Orient the board so that the components face outward and the voltage regulator's heatsink assembly is at the top.

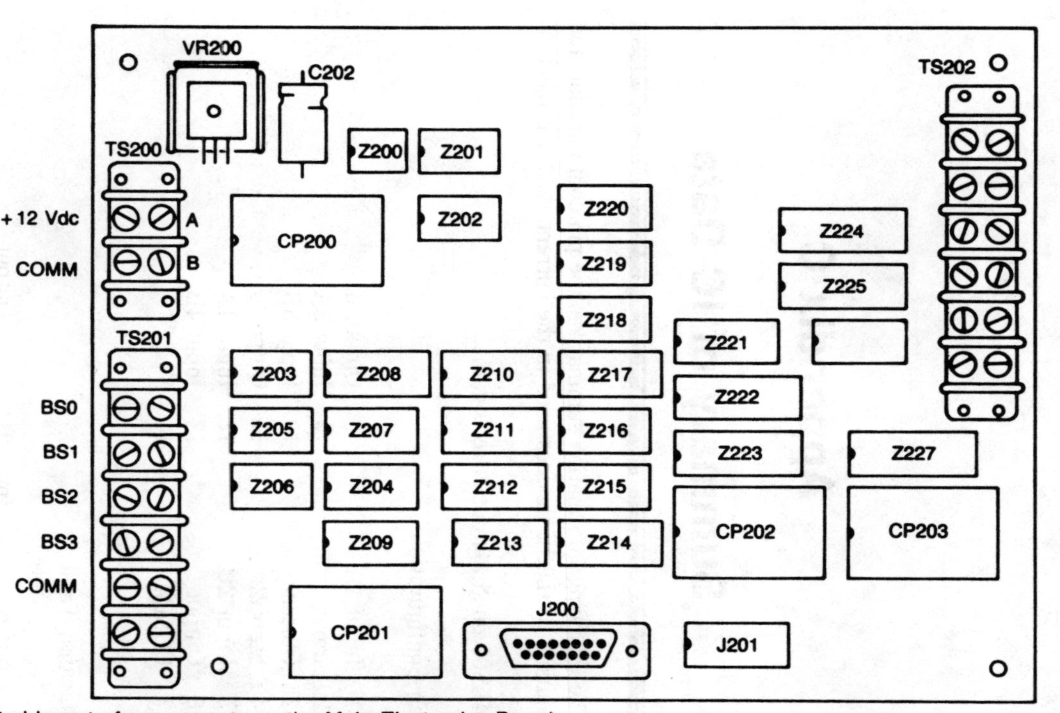

Fig. B-1. Suggested layout of components on the Main Electronics Board.

Appendix C

Summary of IC Data

The following information summarizes the pin configuration and function of the IC devices used in the current Buster project.

74LS00 quad 2-input NAND gate

Pin Configuration

1	Input 1A	8	Output 3
2	Input 1B	9	Input 3A
3	Output 1	10	Input 3B
4	Input 2A	11	Output 4
5	Input 2B	12	Input 4A
6	Output 2	13	Input 4B
7	COMM	14	Vcc

Function Table

Input A	Input B	Output
0	0	1
0	1	1
1	0	1
1	1	0

74LS04 hex inverter

Pin configuration

1	Input 1	8	Output 4
2	Output 1	9	Input 4
3	Input 2	10	Output 5
4	Output 2	11	Input 5
5	Input 3	12	Output 6
6	Output 3	13	Input 6
7	COMM	14	Vc

Function Table

Input	Output
0	1
1	0

7406 hex open-collector inverter

Pin Configuration

1	Input 1	8	Output 4
2	Output 1	9	Input 4
3	Input 2	10	Output 5
4	Output 2	11	Input 5
5	Input 3	12	Output 6
6	Output 3	13	Input 6
7	COMM	14	Vcc

Function Table

Input	Output
0	1
1	0

74LS40 dual 4-input NAND gate

Pin Configuration

1	Input 1A	8	Output 2

2	Input 1B	9	Input 2A
3	not used	10	Input 2B
4	Input 1C	11	not used
5	Input 1D	12	Input 2C
6	Output 1	13	Input 2D
7	COMM	14	Vcc

Function Table

Input A	B	C	D	Output	A	B	C	D	Output
0	0	0	0	1	1	0	0	0	1
0	0	0	1	1	1	0	0	1	1
0	0	1	0	1	1	0	1	0	1
0	0	1	1	1	1	0	1	1	1
0	1	0	0	1	1	1	0	0	1
0	1	0	1	1	1	1	0	1	1
0	1	1	0	1	1	1	1	0	1
0	1	1	1	1	1	1	1	1	0

74LS75 quad latch

Pin Configuration

1	Output 1	9	Output 4
2	Input 1	10	Output 3
3	Input 2	11	Inverted output
4	Enable 3 and 4	12	COMM
5	Vcc	13	Enable 1 and 2
6	Input 3	14	Inverted output 2
7	Input 4	15	Output 2
8	Inverted output 4	16	Output 1

Function Table

Input	Enable	Output	Inverted Output
0	1	0	1
1	1	1	0
X	0	Q	$\overline{Q}$

X = don't care
Q = output prior Enable going to 0
$\overline{Q}$ = inverted Q

74LS93 four-bit binary counter

Pin Configuration

1	Input B	8	Output C
2	Reset 1	9	Output B
3	Reset 2	10	COMM
4	not used	11	Output D
5	VCC	12	Output A
6	not used	13	not used
7	not used	14	Input B

Function Table

Reset		Output			
1	2	D	C	B	A
1	1	0	0	0	0
0	X	count			
X	0	count			

74LS147 or 74147 9-line to 4-line priority encoder

Pin Configuration

1	Input 4	9	Output A
2	Input 5	10	Input 9
3	Input 6	11	Input 1
4	Input 7	12	Input 2
5	Input 8	13	Input 3
6	Output C	14	Output D
7	Output B	15	not used
8	COMM	16	VCC

Function Table

Input									Output			
1	2	3	4	5	6	7	8	9	D	C	B	A
1	1	1	1	1	1	1	1	1	1	1	1	1
X	X	X	X	X	X	X	X	0	0	1	1	0
X	X	X	X	X	X	X	0	1	0	1	1	1
X	X	X	X	X	X	0	1	1	1	0	0	0
X	X	X	X	X	0	1	1	1	1	0	0	1
X	X	X	X	0	1	1	1	1	1	0	1	0
X	X	X	0	1	1	1	1	1	1	0	1	1
X	X	0	1	1	1	1	1	1	1	1	0	0
X	0	1	1	1	1	1	1	1	1	1	0	1
0	1	1	1	1	1	1	1	1	1	1	1	0

X = don't care

74LS151 8-line to 1-line data multiplexer

Pin Configuration

1	Input 3	9	Select C
2	Input 2	10	Select B
3	Input 1	11	Select A
4	Input 0	12	Input 7
5	Output Y	13	Input 6
6	Output W	14	Input 5
7	Strobe	15	Input 4
8	COMM	16	VCC

Function Table

Select			Strobe	Output	
C	B	A	S	Y	W
X	X	X	1	0	1
0	0	0	0	D0	$\overline{D0}$
0	0	1	0	D1	$\overline{D1}$
0	1	0	0	D2	$\overline{D2}$
0	1	1	0	D3	$\overline{D3}$
1	0	0	0	D4	$\overline{D4}$
1	0	1	0	D5	$\overline{D5}$
1	1	0	0	D6	$\overline{D6}$
1	1	1	0	D7	$\overline{D7}$

X = don't care

Dn- = logic level of Input n

$\overline{D}$n- = inverted logic level of Input n

74LS153 dual 4-line to 1-line data multiplexer

Pin Configuration

1	Strobe 1G	9	Output 2Y
2	Select B	10	Data 2A
3	Data 1D	11	Data 2B
4	Data 1C	12	Data 2C
5	Data 1B	13	Data 2D
6	Data 1A	14	Select A
7	Output 1Y	15	Strobe 2G
8	COMM	16	Vcc

Function Table

Select		Strobe	Output
B	A	G	Y
X	X	1	0
0	0	0	DA
0	1	0	DB
1	0	0	DC
1	1	0	DD

X = don't care

Dn- = logic level at Data n

74LS340 8-bit, 3-state inverting buffer

Pin Configuration

1	Strobe A	11	Input B1
2	Input A1	12	Output A4
3	Output B4	13	Input B2
4	Input A2	14	Output A3
5	Output B3	15	Input B3

6	Input A3	16	Output A2
7	Output B2	17	Input B4
8	Input A4	18	Output A1
9	Output B1	19	Strobe B
10	COMM	20	Vcc

Function Table

Input	Strobe	Output
0	0	1
1	0	0
X	1	Hi-Z

X = don't care
Z = high-impedance state

74LS344 8-bit, 3-state noninverting buffer

Pin Configuration

1	Strobe A	11	Input B1
2	Input A1	12	Output A4
3	Output B4	13	Input B2
4	Input A2	14	Output A3
5	Output B3	15	Input B3
6	Input A3	16	Output A2
7	Output B2	17	Input B4
8	Input A4	18	Output A1
9	Output B1	19	Strobe B
10	COMM	20	Vcc

Function Table

Input	Strobe	Output
0	0	0
1	0	1
X	1	Hi-Z

X = don't care
Z = high-impedance state

Index

Edited by David Gauthier